AF855988

interactive SCIENCE

Inquiry Skill Activities
Book 3

PEARSON

Boston, Massachusetts Chandler, Arizona Glenview, Illinois Upper Saddle River, New Jersey

Note to Teachers

This book is designed to give you the greatest flexibility in teaching, reinforcing, and assessing science skills.

Useful at Any Time of the Year

You can teach the skill activities in this book at the beginning of the year or as the course progresses. Students do not need to know formal science concepts or scientific terminology to carry out these activities.

No Lab Equipment Required

All activities are "minds-on." Some simple tools such as rulers, protractors, and compasses are needed for certain skill activities. Graph paper is required for some activities.

Flexible Class Management

These skill activities can be used with the whole class, for small group work, or by individuals. Students could begin working individually in class or at home, then discuss their strategies and results in small groups or as a class.

Copyright © Pearson Education, Inc., or its affiliates. All Rights Reserved. Printed in the United States of America. This publication is protected by copyright, and permission should be obtained from the publisher prior to any prohibited reproduction, storage in a retrieval system, or transmission in any form or by any means, electronic, mechanical, photocopying, recording, or likewise. The publisher hereby grants permission to reproduce these pages, in part or in whole, for classroom use only, the number not to exceed the number of students in each class. Notice of copyright must appear on all copies. For information regarding permissions, write to Pearson Curriculum Group Rights & Permissions, One Lake Street, Upper Saddle River, New Jersey 07458.

Pearson, Prentice Hall, and Pearson Prentice Hall are trademarks, in the U.S. and/or other countries, of Pearson Education, Inc., or its affiliates.

ISBN-13: 978-0-13-369851-0
ISBN-10: 0-13-369851-3

PEARSON

9 10 V001 16 15

Contents

Copyright © Pearson Education, Inc., or its affiliates. All Rights Reserved.

Data Tables and Graphs

Copyright © Pearson Education, Inc., or its affiliates. All Rights Reserved.

Name ______________________ Date ____________ Class __________

INQUIRY SKILL FOCUS Introduction

Observe

People often talk about food—what they like best or which foods are healthful. Food choices often start with observations. You can use observations to gather information about various food choices. When you taste, smell, or see food, you are making observations.

Observing is using one or more of your senses—sight, hearing, smell, taste, and touch—to gather information about the world. The way food smells and tastes are observations. The color and shape of a piece of fruit are observations as well. Observing is a skill you use every day. It's an important tool in scientific investigations as well.

Observations in science must be accurate. Accurate observations are those that are carefully made and correctly recorded. Observations in science must also be objective. An objective observation avoids bias, or a certain point of view. What are some things you might say about a particular food?

Example 1: This cereal contains raisins. (accurate and objective)

Example 2: All healthy cereal contains raisins. (not accurate)

Example 3: Cereal with raisins is delicious. (not objective)

Two kinds of observations are used in science—qualitative observations and quantitative observations. Qualitative observations do not use numbers. For example, if you see that certain types of crackers are round, you are making a qualitative observation. Quantitative observations are observations that include numbers. "There are 65 crackers in this box" is a quantitative observation. Tools are often used to make quantitative observations.

Example 4: This orange juice contains pulp. (qualitative)

Example 5: This grape juice is purple. (qualitative)

Example 6: There are 250 mL of juice in the glass. (quantitative)

Example 7: Fifteen students in science class had juice at breakfast. (quantitative)

Copyright © Pearson Education, Inc., or its affiliates. All Rights Reserved.

Name ______________________ Date ______________ Class __________

Observe *(continued)*

In science, observations are often used to make inferences, or explanations for what was observed. Inferences are different than observations. Accurate observations are correct, factual statements. Inferences are not always correct. Be sure your observations include only information you gather with your senses.

Example 8: I smelled fresh bread when I walked in the school hallway this morning. (observation)

Example 9: More than 200 students are in the cafeteria right now. (observation)

Example 10: Today's school lunch will probably be pizza. (an inference, not an observation)

Example 11: Pizza is most students' favorite lunch. (an inference, not an observation)

TIPS FOR MAKING OBSERVATIONS

- Qualitative observations are made by using your senses. Remember—never taste, touch, or smell a chemical or substance unless your teacher tells you to.
- Review your observations. Make sure they are accurate and objective.
- Quantitative observations are often made using tools. Meter sticks, balances, graduated cylinders, and thermometers are examples of tools used in science. Include units with all measurements. Quantitative observations are sometimes made by counting.
- If tools aren't available, use estimation to make quantitative observations. Comparing to an everyday object is one way to estimate. You could say that an object is about as wide as a penny or has about the same mass as a pencil.
- Review your observations to be sure they are not inferences. Remember, observing means using your senses to gather information. Inferences are possible explanations for your observations.

Checkpoint Write three observations about your desk or other furniture in the classroom. Label each observation as qualitative or quantitative. Make sure your statements are not inferences.

____

____

____

____

Copyright © Pearson Education, Inc., or its affiliates. All Rights Reserved.

INQUIRY SKILL FOCUS Practice

Observe

Use the illustration to answer the questions that follow. Write your answers on a separate sheet of paper.

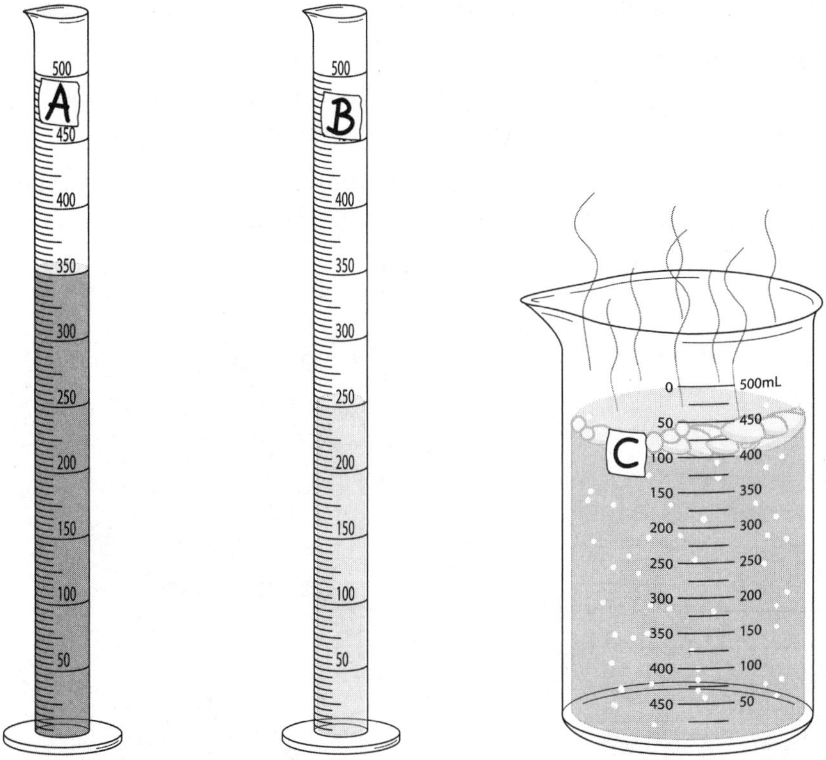

❶ Make three qualitative observations about the illustration.

❷ Make three quantitative observations about the illustration.

For each statement below, tell whether the statement is an observation or an inference. Explain your answer.

❸ When Liquid A and Liquid B are mixed, there is a chemical reaction.

❹ Liquid A is a different color than Liquid B.

❺ Liquid C was made by combining 150 mL of Liquid A and 250 mL of Liquid B.

❻ **Think It Over** Write a few sentences explaining how you know whether a statement is an observation or an inference. Include an example as part of your answer.

____

____

____

Copyright © Pearson Education, Inc., or its affiliates. All Rights Reserved.

Name ______________________ Date ____________ Class ________

INQUIRY SKILL FOCUS Introduction

Infer

Have you ever woken during the night to the sound of raindrops hitting your window? Right away you think, "It's raining outside." You used an observation—the sound of rain hitting the window—and your past experience to form a conclusion about the weather. This kind of conclusion is called an inference.

Inferring is explaining, or interpreting, your observations. Reasonable inferences make sense, and are based on observations and prior knowledge. Unreasonable inferences may conclude too much from an observation.

Suppose you walk into a neighbor's yard and observe the scene shown below. What are some observations and inferences you could make?

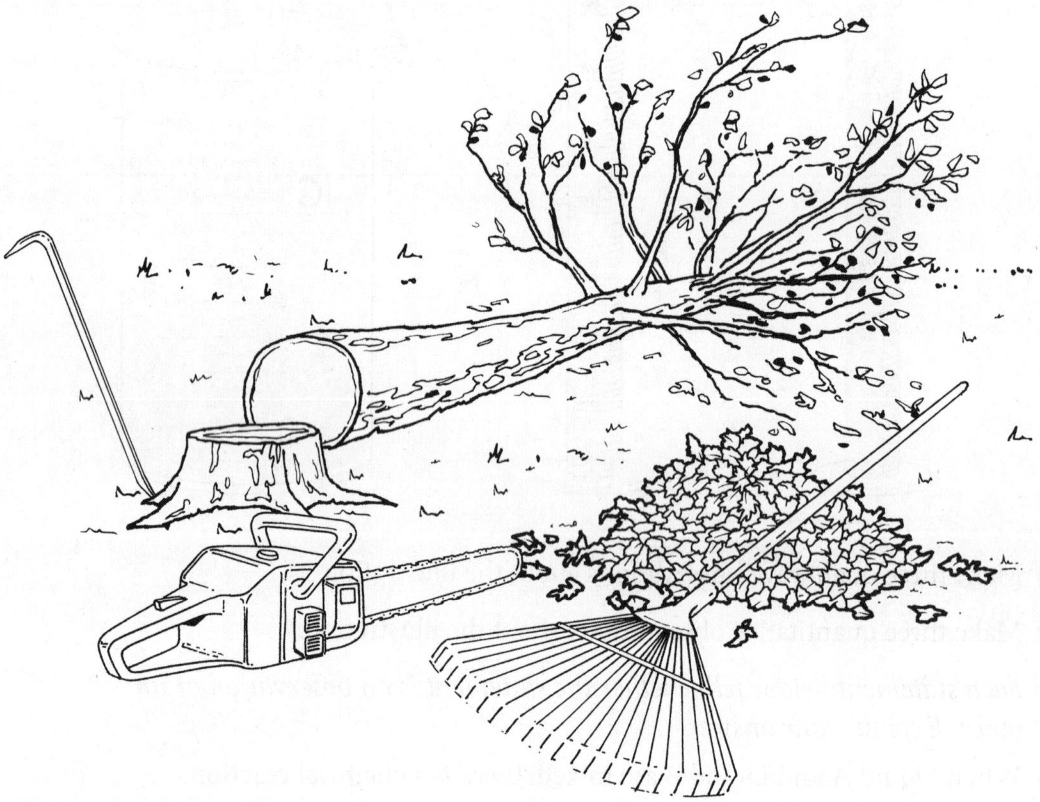

Observation: The tree is lying on the ground.

Inference 1: The saw was used to cut down the tree. (reasonable)

Inference 2: This saw has been used to cut down many trees. (unreasonable, because you have no information about how often the saw is used)

Observation: The leaves are in a pile.

Inference 1: The rake was used to move the leaves into a pile. (reasonable)

Inference 2: Two people worked together to rake the leaves. (unreasonable, because your observation gives no information about the number of people who worked in the yard)

Copyright © Pearson Education, Inc., or its affiliates. All Rights Reserved.

Name ______________________ Date ____________ Class __________

Infer *(continued)*

You can often use one observation to make several inferences. For example, if you wake up and hear water drops hitting the window, you can infer that it is raining. What other inferences might you make using this same observation?

Observation: I hear water hitting the window.

Inference 1: The grass is wet. (reasonable)

Inference 2: Today's baseball game might be canceled. (reasonable)

Each of the inferences above is reasonable based on the observation. Reasonable inferences are not always correct, though. After you've made inferences, you can determine if they are correct. In the examples above, you could touch the grass to see if it is wet. You could call the coach to see if the baseball game is canceled. In science, inferences are often verified by observations or with experiments.

TIPS FOR MAKING AN INFERENCE

- Be sure your observations are correct before using them to make inferences.
- Use prior knowledge and past experiences when making inferences.
- Try to make more than one logical inference from this same observation.
- Review each of your inferences. Make sure each inference is reasonable and logical.
- Identify what information you need to determine whether your inference is correct or incorrect.
- Gather more information and check your inferences for accuracy.

Checkpoint On a family trip, you observe a car by the side of the road. There is a jack under the car, and one tire has been removed. What is one inference that you could make from this observation? Can you be sure your inference is correct? Why or why not?

____

____

____

____

____

Copyright © Pearson Education, Inc., or its affiliates. All Rights Reserved.

Name ______________________ Date ______________ Class __________

INQUIRY SKILL FOCUS Practice

Infer

The diagram below shows the beaks of several different species of birds. Make observations about the beaks and then answer the following questions. Write your answers on a separate sheet of paper.

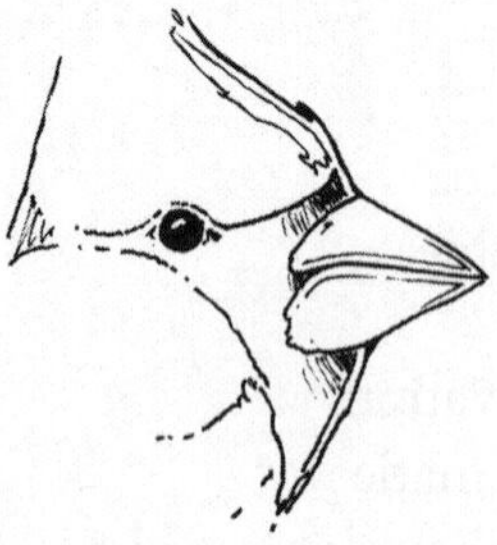

1.

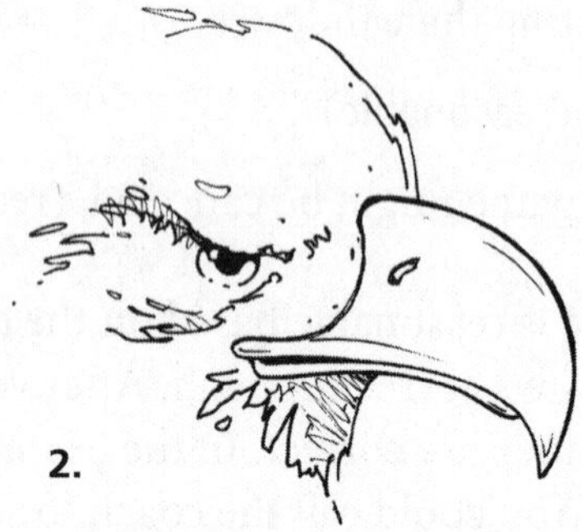

2.

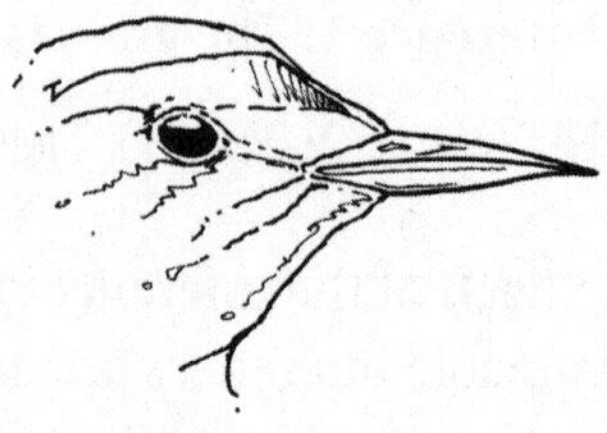

3.

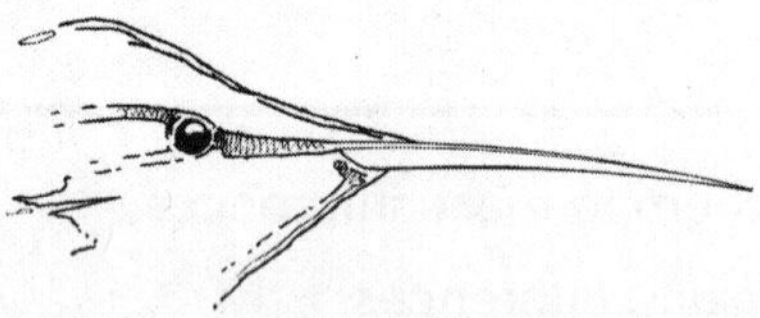

4.

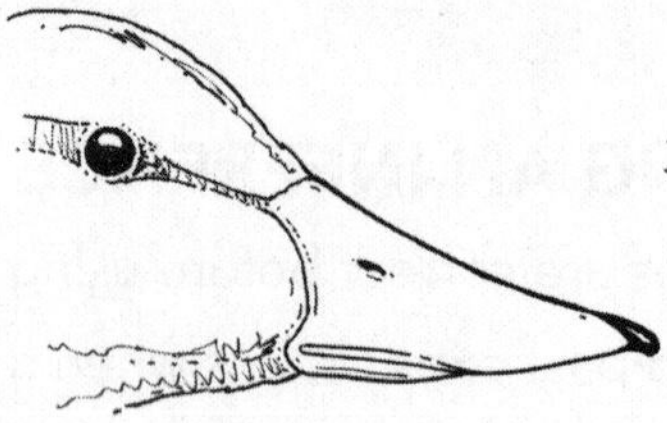

5.

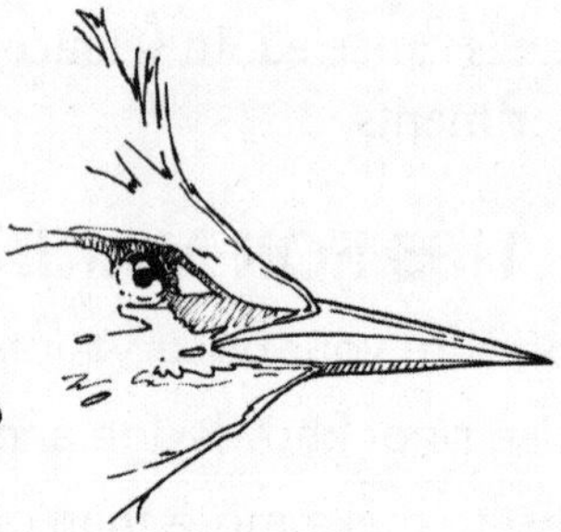

6.

1. The seeds that make up the diet of some birds are difficult to crack. A thick beak is needed to crack the seeds. Most seed-eaters have short, thick beaks in comparison to other birds. Use observations and inferences to determine which of the birds shown here could be a seed-eater.
2. Some birds gather nectar from the inner parts of flowers. Use inferences to describe a beak that would be well-suited for getting nectar from deep within flowers.
3. Which bird or birds in the diagram above do you infer eat nectar?
4. A strong, curved or hooked beak is used by some birds to catch prey and tear meat. Make an inference about which bird in the diagram eats meat.

Copyright © Pearson Education, Inc., or its affiliates. All Rights Reserved.

Name ______________________ Date ____________ Class __________

Infer *(continued)*

You can make inferences about a bird's habitat and diet by observing the structure of the bird's feet. The diagram below shows the feet of several different species of birds. Make observations about the feet, and then answer the following questions. Write your answers on a separate sheet of paper.

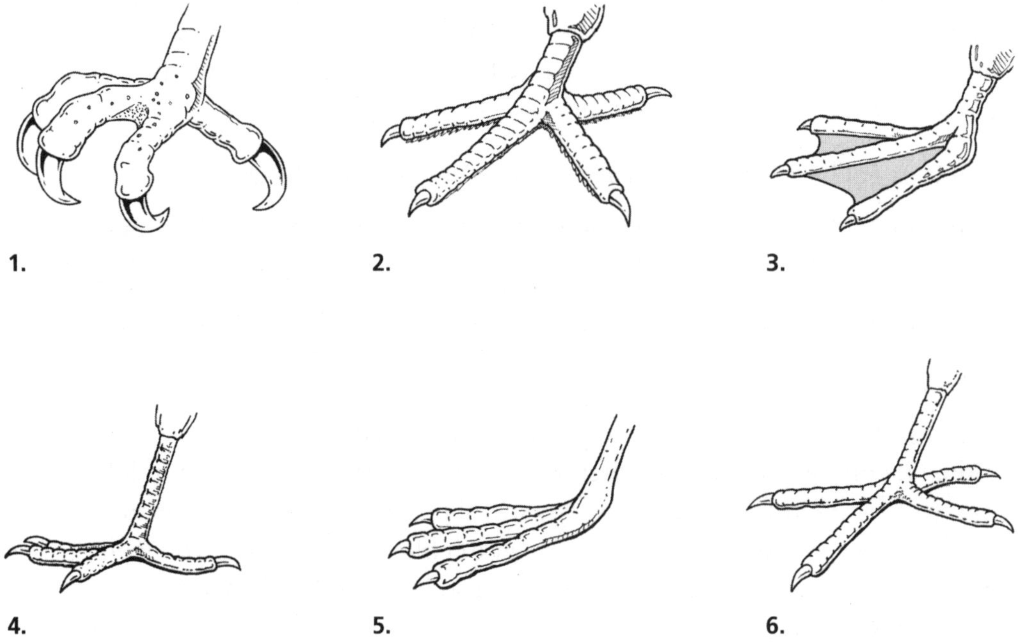

5. Some birds grasp their prey. Make an inference about which of the bird's feet shown here would be best for grasping prey.

6. A fourth toe makes a bird run more slowly. Which bird or birds in the diagram do you infer are particularly fast runners?

7. Infer which bird foot or feet belongs to a bird that swims.

8. What observation did you use to make your inference to answer Question 7?

9. **Think It Over** What kind of information would allow you to determine if your inferences about birds are correct? What are two different ways you could get this information?

____

____

____

Copyright © Pearson Education, Inc., or its affiliates. All Rights Reserved.

Name ______________________ Date ____________ Class __________

INQUIRY SKILL FOCUS Introduction

Predict

Each of the first six weeks of the school year, your science teacher has given a surprise quiz on Friday. After you notice the pattern, you begin to study each Thursday night to prepare. You predict there will be a quiz on every Friday. You often make predictions in your everyday life. Predictions are an important skill in science as well.

A **prediction** is an inference about a future event based on evidence and past experience. Some predictions are very specific. Other predictions are general, or less detailed. You might ask a friend to predict how well he will do on the math test this week. He could answer in several different ways.

Prediction 1: I'll probably get a good grade on my math test. (general)

Prediction 2: I'll probably score higher than 80% on my math test. (somewhat specific)

Prediction 3: I probably won't miss more than one problem on the math test this week. (very specific)

Patterns, or trends, in data can help you to make specific, accurate predictions. If you were asked to predict the high temperature for the 4th of July next year, you could probably make a very general prediction based on previous knowledge. However, if you could examine a graph like the one below, you would be able to make a more specific prediction. It is also more likely that your prediction would be accurate.

Notice that you might make different predictions based on these data. For example, during an unusually hot year you might make Prediction 6.

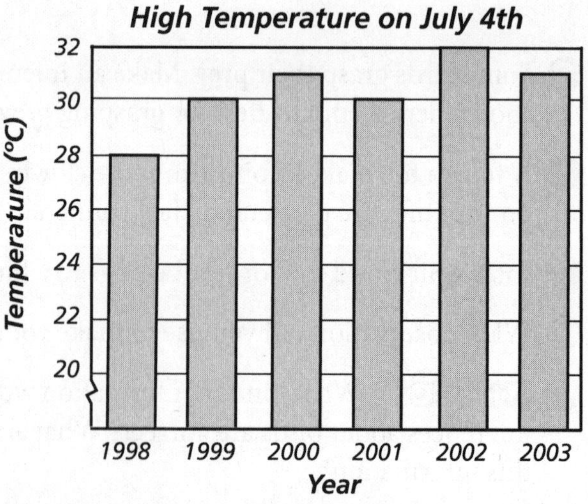

Prediction 4: The high temperature will be about 30°C. (matches the highest temperatures given in graph)

Prediction 5: The high temperature will be 32°C. (matches the highest temperature in the graph)

Prediction 6: The high temperature will be 33°C. (exceeds the highest temperature in the graph)

Copyright © Pearson Education, Inc., or its affiliates. All Rights Reserved.

Name ______________________ Date ____________ Class ________

Predict *(continued)*

Not all predictions are correct. You might predict a certain temperature for the 4th of July based on all of the information available, but your prediction might not be correct. A prediction is only an inference about something that will happen in the future, and as you've learned, inferences aren't always correct.

You can use observation or investigation to check your predictions. Many predictions in science must be checked using a controlled experiment. If a scientist predicted that increasing the temperature of a solvent would cause more of a substance to dissolve in it, she could run a controlled experiment to test her prediction. Testing predictions is one way that scientists increase their knowledge of the natural world.

TIPS FOR MAKING PREDICTIONS

- Use observations and your past experiences to make predictions.
- Patterns, or trends, in data can help you make specific, accurate predictions.
- Tables, charts, and graphs can help you see patterns in data.
- Make your predictions as specific as possible.
- Use background information when you make a prediction. Don't just make a guess.
- Controlled experiments and observations can be used to check the accuracy of your predictions. If you check your prediction and find is not accurate, you still have learned something.

Checkpoint Write an example of a prediction that you could test using a controlled experiment. Write a few sentences describing how you could test this prediction.

____

____

____

____

____

____

____

____

____

____

Copyright © Pearson Education, Inc., or its affiliates. All Rights Reserved.

Name ______________________ Date ______________ Class __________

INQUIRY SKILL FOCUS Practice

Predict

If you've ever made a recipe with vinegar, you've used a solution. A solution has two parts. Vinegar is a solution of water and acetic acid. The solvent is the part of the solution present in the largest amount. The solute is the substance dissolved by the solvent. In the example, water is the solvent. Acetic acid is the solute. Some substances dissolve more readily than others. The amount of solute that will dissolve is also affected by the temperature of the solvent.

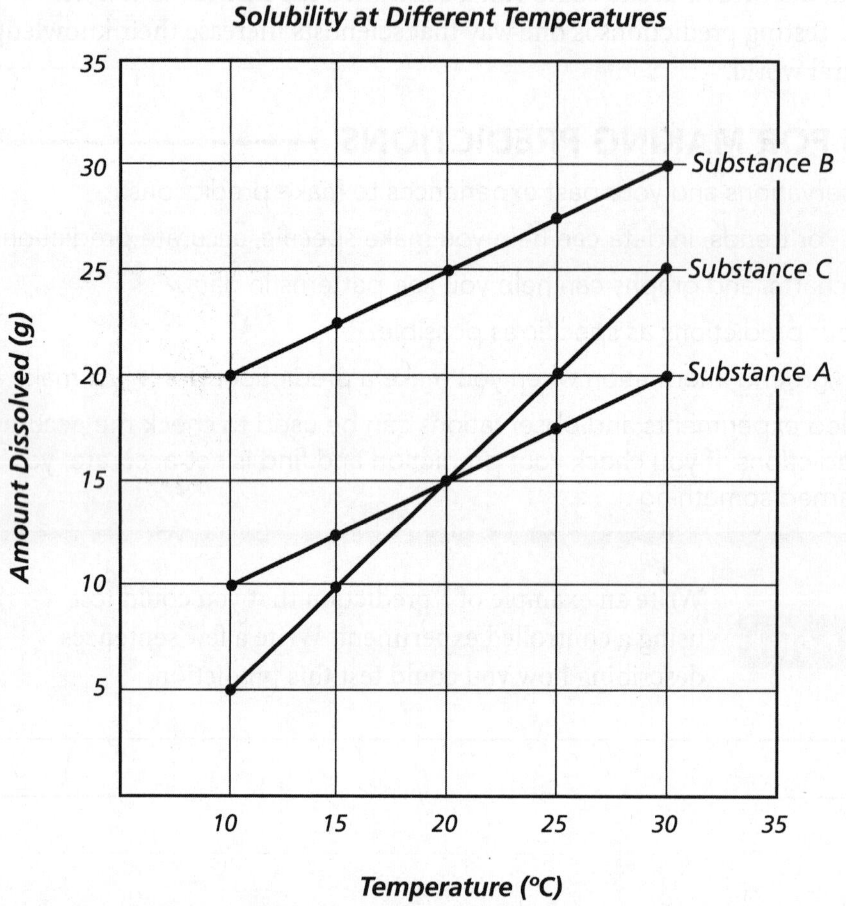

Copyright © Pearson Education, Inc., or its affiliates. All Rights Reserved.

Name ______________________ Date ____________ Class ________

Predict *(continued)*

Use the graph on page 10 to answer the questions below. Write your answers on a separate sheet of paper.

The graph on page 10 shows the amount of three different substances (solutes) that will dissolve in a another substance (solvent) at different temperatures. The graph shows how the temperature of the solvent affects the solubility of each solute. This graph can be used to make predictions about the solubility of these substances at other temperatures.

1. Describe two trends, or patterns, you see in the data shown in the graph.
2. Make a prediction to answer the question. Could you dissolve more of Substance A or Substance B in this solvent at 35°C?
3. Explain how you made your prediction for Question 2.
4. Examine the graph lines for Substance A and Substance C, including the point where they intersect. Then predict whether Substances A and C will have equal solubilities at temperatures higher than 20° C.
5. Explain how you made your prediction for Question 4.
6. **Think It Over** Based on the graph, can you be sure your predictions in Questions 2 and 4 are correct? Explain the procedure you would use to test the prediction you made in Question 2.

Copyright © Pearson Education, Inc., or its affiliates. All Rights Reserved.

Classify

Imagine walking into the grocery store to buy the ingredients for your favorite meal, only to find that the food on the store shelves was not organized in any way. To find a particular item, you have to search the entire store! Grocery stores organize, or classify, food products so you can easily find the things you need.

Classifying is organizing objects and events into groups according to a system, or organizing idea. Objects can be classified by size, shape, properties, or use. Scientists also classify living things in many different ways.

Classification makes finding information easier and shows the relationship between the items being classified. There is usually more than one correct way to classify objects. Rocks, for example, can be grouped by their texture, color, mineral composition, or origin. The classification scheme and organizing ideas used depend on the purpose of the classification. A scientist studying the history of Earth might chose to classify rocks based on their origin. A student interested in making an attractive rock display might choose to classify rocks by their color.

Classification systems start by dividing the whole group into parts, based on one organizing idea. Some classification systems start with objects being divided into two groups. Other systems use three or more starting groups. This diagram shows how rocks can be classified into three groups. The organizing idea for this classification system is how the rocks were formed.

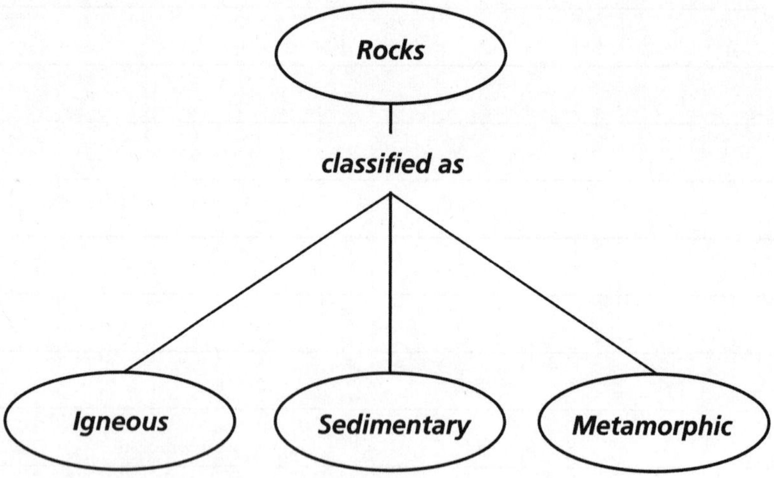

Copyright © Pearson Education, Inc., or its affiliates. All Rights Reserved.

Classify *(continued)*

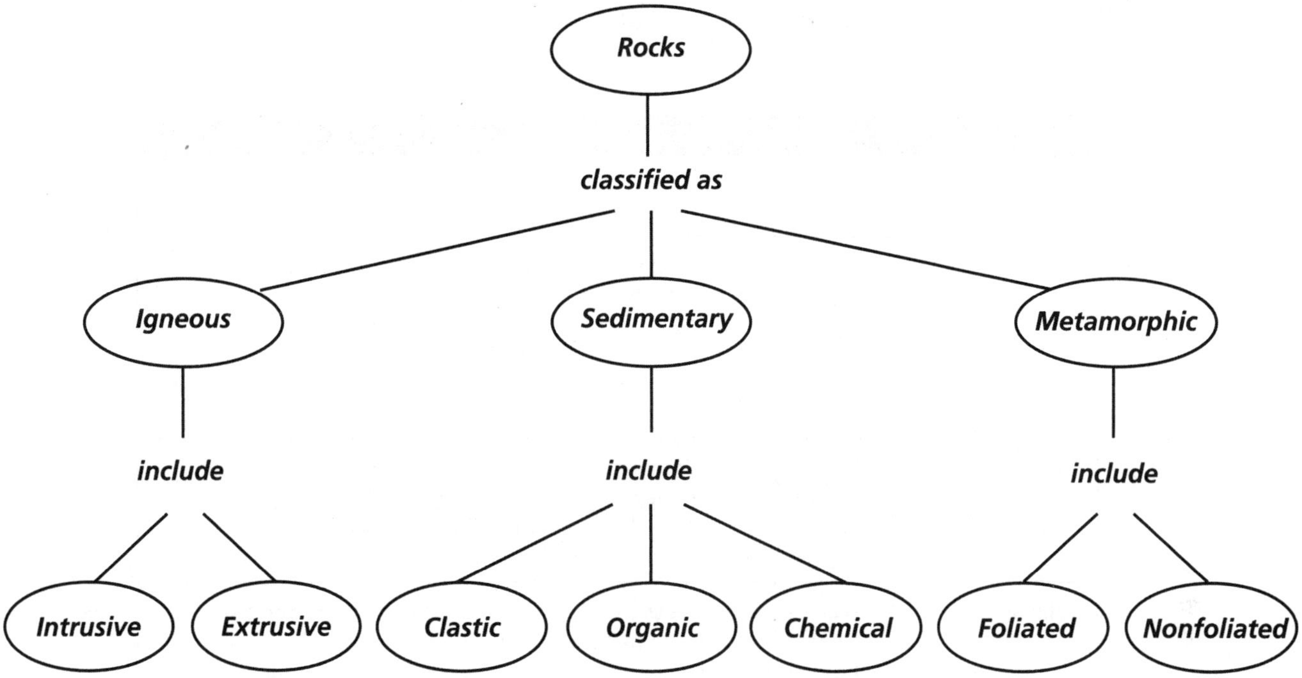

These groups can be further classified using different organizing ideas. The first level of classification divides the rocks into three groups based on their origin. The second level uses other organizing ideas to further classify the rocks.

TIPS FOR CLASSIFYING

- Observe the objects you are classifying. Decide which similarities and differences can be used for classification.
- Place the objects into groups based on your organizing idea.
- Decide if the groups can be broken down into smaller groups. Use another of the similarities or differences to further classify the objects.

Copyright © Pearson Education, Inc., or its affiliates. All Rights Reserved.

Name ______________________________ Date ______________ Class ___________

INQUIRY SKILL FOCUS Practice

Classify

Use this table to answer the questions on page 15.

Energy Resources	
Type of Resource	**Description**
Coal	This fossil fuel is formed from plant remains over millions of years. Coal is mined in locations throughout the United States and the world. Coal supplies 23% of the energy used in the United States. Because coal takes millions of years to form, it is considered a nonrenewable resource.
Oil	Oil is a fossil fuel formed from the remains of plants. Oil is widely used for energy in the United States, but only 3% of the oil in the world is located in the United States. Oil takes millions of years to form, so it is considered a nonrenewable resource.
Solar energy	Solar energy is energy from the sun. It can be captured and used to heat homes and to make electricity in solar power plants. Solar energy is available at most locations, but only when the sun is shining. For this reason, and because converting solar energy to electricity is expensive, it is not widely used in the United States. Solar energy is considered an inexhaustible resource.
Tidal energy	Tides provide a source of energy that can be converted to electricity. However, very few locations in the world are good locations for tidal power plants. Tidal power supplies very little of the world's electricity. Although there is an unlimited supply of tidal energy, it is very difficult to use.
Wind energy	Wind is an inexhaustible source of energy. Wind is not widely used as an energy source. In fact, less than 1% of the world's electricity is generated using wind energy. Few places have winds that are steady enough to be good locations for collecting wind energy. Wind energy does not create pollution.
Wood	Wood can be burned as a source of energy. Wood is a renewable resource, because trees can be replaced in a reasonable amount of time. Wood is found throughout the United States and the world. Wood is not widely used as a source of energy in the United States.

Copyright © Pearson Education, Inc., or its affiliates. All Rights Reserved.

Classify (continued)

Use the information on the previous page to answer the following questions. Write your answers on a separate sheet of paper.

1. Describe three different organizing ideas that could be used to divide the energy resources into groups.

2. Describe an organizing idea that you could use to divide the energy sources into three groups.

3. Diagram a two-level classification of the energy resources. Be sure to write the organizing ideas you used to make each group. You can use the organizing ideas from Questions 1 and 2, or develop new organizing ideas. List the energy resources in each group.

4. **Think It Over** Imagine you are a journalist assigned to write about energy production and its effect on the environment. What are two organizing ideas that you could use to classify energy sources for your article? Use your general knowledge as well as the information on page 14.

Copyright © Pearson Education, Inc., or its affiliates. All Rights Reserved.

Name ______________________ Date ______________ Class __________

INQUIRY SKILL FOCUS Introduction

Make Models

A globe is a model used to show locations on Earth. Scientific **models** are pictures, diagrams, or other representations of objects or processes.

A chemical equation is an example of a mental model—a model that shows a process. This chemical equation shows how hydrogen and oxygen combine to make water. The chemical equation is one example of a model. The same process can be modeled using pictures of the atoms and molecules.

$$H_2 + ½ O_2 \rightarrow H_2O$$

Hydrogen *Oxygen* *Water*

Some physical models, such as maps, are two-dimensional. Other physical models, such as globes, are three-dimensional. Some models used in science are scale models. A scale model shows how large the parts of an object are in relation to one another and to the whole object. A model might use the scale 1 cm = 1 m. That means a part of the model that is 1 centimeter long represents an actual object that is 1 meter long. Some scale models, like a scale model of a blood cell, are larger than the actual object. Other scale models, like a model of the solar system, are smaller than the object being studied.

TIPS FOR MAKING MODELS

- Think about the object or process to be modeled. Decide whether a diagram, picture, or three-dimensional model would best represent the object or process.
- To model a process, think through the entire process before you begin. Write down the steps of the process in order.
- To make a physical model, obtain all the needed materials before you begin.
- Decide whether the model will be larger than, smaller than, or the same size as the real object.
- Decide whether your model will be a scale model. Determine the scale you will use before you begin making your model.

Checkpoint In the water cycle, water evaporates from Earth's surface into the atmosphere. It later returns to the surface as precipitation. How would you make a model of the water cycle? Write a paragraph describing the model. Could you make a scale model? Why or why not?

____

Copyright © Pearson Education, Inc., or its affiliates. All Rights Reserved.

Name ______________________ Date ____________ Class ________

INQUIRY SKILL FOCUS Practice

Make Models

The paragraph and table below give information about Earth's atmosphere. Use that information to plan a two-dimensional scale model of some of the following layers of Earth's atmosphere: troposphere, stratosphere, mesosphere, and ionosphere. Use the graph paper on page 18 to make your model.

Earth's atmosphere is divided into layers based on changes in temperature. The troposphere is the lowest layer of Earth's atmosphere. This is the layer in which weather occurs, and it's the layer in which you live. The stratosphere contains ozone, which absorbs ultraviolet radiation from the sun. The mesosphere is considered the middle layer of the atmosphere, and it's also the coldest layer. The thermosphere is divided into two layers. The ionosphere is the lower layer of the thermosphere. Temperatures in the ionosphere can reach 1,800°C. The exosphere is the outer layer of the thermosphere. It has no definite outer limit.

Earth's Atmosphere

Layer		Location
Troposphere		From Earth's surface to 12 km above the Earth's surface
Stratosphere		From 12 km to 50 km above Earth's surface
Mesosphere		From 50 km to 80 km above Earth's surface
Thermosphere	Ionosphere	From 80 km to 550 km above Earth's surface
	Exosphere	Starts 550 km above Earth's surface, and has no definite upper limit

Write the answers to the following questions on a separate sheet of paper.

1. What scale will you use for your model? Why?
2. Using the scale you have chosen, determine the final size of your model.
3. **Think It Over** Describe some of the difficulties you experienced making your model. How did you solve them? Include an explanation of why the exosphere can't be included in a scale model.

Copyright © Pearson Education, Inc., or its affiliates. All Rights Reserved.

Name ______________________ Date ____________ Class ________

Make Models *(continued)*

Use the space below to sketch your model of Earth's atmosphere. Include the scale you need for your model.

Copyright © Pearson Education, Inc., or its affiliates. All Rights Reserved.

Name ______________________________ Date ______________ Class __________

INQUIRY SKILL FOCUS Introduction

Communicate

When you write a letter, talk with a teacher, or chat with your friends, you are sharing your ideas with other people. Communicating is the process of sharing ideas and information with others. Communicating is something you do every day. Scientists also communicate as a part of their job. Scientific communication allows scientists to share ideas and repeat the work of others.

Scientists communicate their results and ideas in papers, in talks, at meetings, and in informal communication such as e-mails and telephone conversations. Communication allows scientists to share ideas for future work, verify the results of others' work, and repeat experiments. Scientific communication follows these guidelines:

- Every step in a procedure should be described completely and in detail. The written procedure should enable someone else to duplicate the work.
- Observations and measurements should be accurately recorded. Even if the experiment doesn't turn out as planned, all results must be recorded.
- Record inferences separate from observations. Be certain that someone reading the results can tell the difference between the observations and the inferences.
- Review the work to be sure it is objective. Consider all of the possible explanations for the work.

Like communication in everyday life, scientific communication can be either oral or written.

Oral Communication Sometimes ideas are shared by talking and listening, or communicating orally. In science, some oral communication is informal. Scientists often use the telephone or meet with others to discuss their investigations and results. Scientists also use oral communication skills to make formal presentations. You will use oral communication when you work with small groups, make presentations to your class, or discuss your results at science fairs and other events. Some of the guidelines for oral communication in science are:

- Take your time when you are discussing ideas. Speak slowly so your audience can understand you. Define any terms that might be unfamiliar to your audience.
- Use tables, charts, graphs, and other visual aids whenever possible. Visual displays can help make complex ideas clear to your audience.

Copyright © Pearson Education, Inc., or its affiliates. All Rights Reserved.

Name ______________________ Date ____________ Class ________

Communicate *(continued)*

Written Communication Scientists submit written reports of their work to scientific journals. The written work is reviewed by other scientists before it is published in the journals. You will also need to submit written work in science. Lab reports are one form of written communication in science. Lab reports usually contain the following information:

1. problem or question
2. hypothesis
3. list of materials
4. procedure
5. observations, often organized in data tables
6. analysis, including any calculations and graphs
7. conclusion

As you work on an investigation, you might have ideas for additional work. You might think of additional questions. If so, these are noted at the end of the lab report.

TIPS FOR COMMUNICATING IN SCIENCE

- Record your observations honestly. Make sure your observations are complete and detailed.
- Use data tables and charts to record your data. This keeps your data neat and organized.
- Write down your procedure. Make sure the procedure is written with enough details for someone else to duplicate the work.
- When you perform mathematical calculations, write down the equations and formulas you use. Make sure to include units with your measurements and calculations.
- Follow any specific rules for a particular project. An informal class report may have different guidelines than a formal written report.
- Make notes if you include the work of others in your reports. Provide the names of your sources.

Checkpoint Name two ways in which oral and written communication are similar. What are some advantages of each type of communication?

____

____

Copyright © Pearson Education, Inc., or its affiliates. All Rights Reserved.

Name ______________________________ Date ______________ Class __________

INQUIRY SKILL FOCUS Practice

Communicate

Lab reports are one form of communication used in science. Use the description below and the questions on the next page to write a lab report for this investigation.

This investigation was about pulse rate. We checked our pulse rates when we were resting and during different kinds of exercise. I thought that exercise would make our pulse rates increase. We used a watch with a second hand to time ourselves when we measured our pulse rates. We measured our pulse for 15 seconds. We can use this measurement to calculate our pulse rates per minute. We also checked our pulse five minutes after exercise. We wanted to see if our pulse rates returned to the resting level.

Pulse rates (15 seconds):

Student 1: resting 19; after one minute of jumping jacks 30; after walking up stairs for one minute 35; 5 minutes after exercise 23

Pulse rates (15 seconds)

Student 2: resting 22; after one minute of jumping jacks 35; after walking up stairs for one minute 40; 5 minutes after exercise 23

We will need to do calculations in order to convert these results to pulse rate per minute. These results should allow us to see the effect of exercise on our pulse rates. While we were investigating, we noticed that there are differences between our pulse rates. We wonder how much variation there is among all of the students in our class.

Copyright © Pearson Education, Inc., or its affiliates. All Rights Reserved.

Name ______________________ Date ____________ Class __________

Communicate *(continued)*

Write your answers to the following questions on a separate sheet of paper.

1. State the problem or question this student was investigating.
2. Use the *If…then…*form to write a hypothesis for this investigation.
3. Read the student's notes carefully and write a complete materials list for the investigation.
4. Write a step-by-step procedure for this investigation. Remember that the procedure should be written so that the work could be repeated exactly.
5. Organize the student's observations in a data table that will neatly and clearly show the data from the investigation.

Student Pulse Rates

	Pulse Rate (beats/15 s)	
Activity	**Student 1**	**Student 2**

6. Data analysis is the next step. Show the calculations required for this lab report. Be sure to use units with your numbers.
7. Provide a conclusion for the investigation.
8. **Think It Over** The students had an additional question while they were performing this investigation. Write down a different question related to the students' work and describe how it could be investigated.

Copyright © Pearson Education, Inc., or its affiliates. All Rights Reserved.

Name ______________________________ Date ______________ Class __________

INQUIRY SKILL FOCUS Introduction

Measure

Can you imagine trying to build a set of wooden bookshelves without using measurements? Measurements are necessary to be sure that the pieces of wood you cut will fit together. Measurements are also necessary to be certain that the finished bookshelf will fit through the door. Measurements are important for any construction project. In science, measurements are also necessary. Standard measurements allow scientists to repeat the work of others.

Measuring is comparing an object or process to a standard. In science, a common set of standards, called the International System of Units, is used for measuring. This system, a version of the metric system, is sometimes called SI (for its French name, *Système International d'Unités*). Common SI units include the meter, the liter, the gram, and degrees Celsius.

Common SI Units

Property	Basic Unit	Symbol
Length	meter	m
Liquid volume	liter	L
Mass	gram	g
Temperature	degrees Celcius	°C

Sometimes units smaller or larger than the standard units are used. A set of prefixes is used to name smaller and larger units and show their size relative to the standard unit. For example, a kilometer is equal to 1,000 meters. The kilometer is useful when measuring long distances. A milligram is 1/1,000 gram, and is used when measuring objects with very small masses. To convert between units, you must multiply or divide. For example, to change a measurement in kilometers to a measurement in meters, you would multiply by 1,000.

Common SI Prefixes

Prefix	Symbol	Meaning	Example
kilo-	k	1,000	kilometer (km)
hecto-	h	100	hectometer (hm)
deka-	da	10	dekameter (dam)
deci-	d	0.1 $\left(\frac{1}{10}\right)$	decimeter (dm)
centi-	c	0.01 $\left(\frac{1}{100}\right)$	centimeter (cm)
milli-	m	0.001 $\left(\frac{1}{1,000}\right)$	millimeter (mm)

Copyright © Pearson Education, Inc., or its affiliates. All Rights Reserved.

Name ______________________ Date ____________ Class __________

Measure *(continued)*

In science, tools are used to make measurements. Selecting the correct tool and using the tool properly are both important steps in making correct measurements. Length can be measured using a centimeter ruler or a meter stick. Liquid volume is often measured using a graduated cylinder. Balances are used to measure mass. Thermometers are used to measure temperature.

TIPS FOR MAKING MEASUREMENTS

- Select the appropriate unit and tool for the measurement you are making.
- Make sure you know how to correctly use the measuring tool. If you have questions, ask your teacher.
- Always label your measurements with the appropriate units.
- If you perform mathematical calculations using your measurements, be sure to label the results with the correct units.

Checkpoint What is 420 kilometers expressed as meters? Show the calculations you used to find your answer.

Copyright © Pearson Education, Inc., or its affiliates. All Rights Reserved.

Name ______________________ Date ______________ Class __________

INQUIRY SKILL FOCUS Practice

Measure: Length

Write your answers to the questions below in the spaces provided.

Length is the distance between two points. The SI unit for length is the meter (m). Meters can be divided into smaller units, such as the centimeter (cm) and millimeter (mm), to measure small objects. Meters are combined into larger units, such as the kilometer (km), to measure long distances.

Common tools used to measure length are meter sticks and centimeter rulers. A centimeter ruler is shown below. The labeled units are centimeters (cm). Each centimeter is divided into 10 millimeters. When measuring length, choose the unit and measuring tool that is most appropriate for the object being studied. If you are comparing the lengths of several objects, choose a unit that can be used to measure all of the objects.

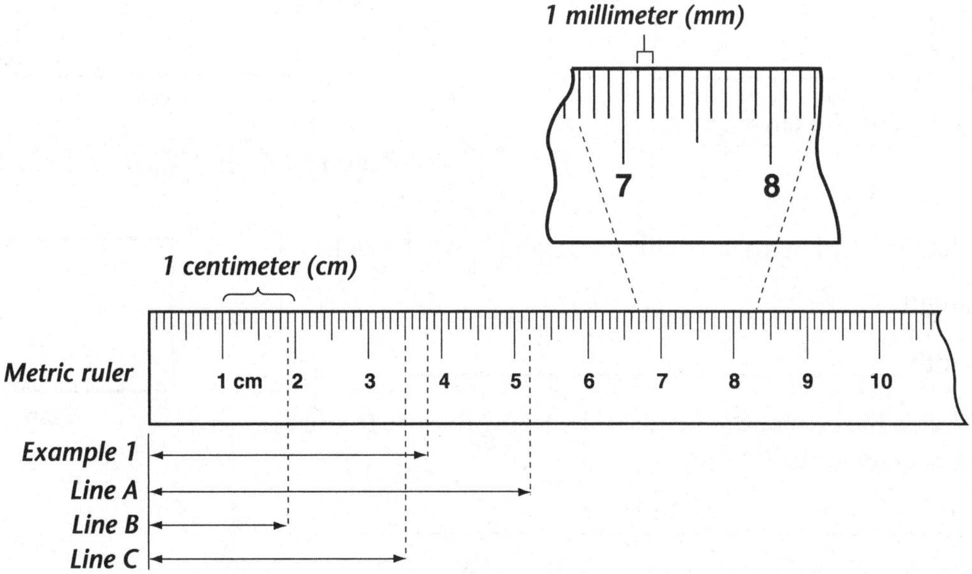

Look at Example 1 above. If you measure this line in centimeters, it is 2.9 cm long. In millimeters, this line is 29 mm long.

1. How many millimeters long is Line A? ______________________
2. How many centimeters long is Line A? ______________________
3. How many millimeters long is Line B? ______________________
4. How many centimeters long is Line B? ______________________
5. How many millimeters long is Line C? ______________________
6. How many centimeters long is Line C? ______________________
7. What is the difference between lines A and B? ______________________ cm

Copyright © Pearson Education, Inc., or its affiliates. All Rights Reserved.

Name ______________________ Date ____________ Class ________

Measure: Length *(continued)*

Using Length Measurements to Find Area and Volume

You can use metric measurements to find the area of a rectangular figure by multiplying length and width. Area is measured in units squared.

You can use metric measurements to find the volume of a solid rectangular figure by multiplying the length, width, and height. Volume is measured in units cubed.

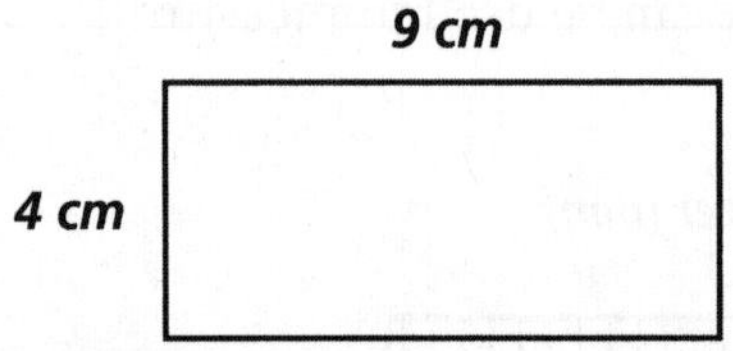

Area = 9 cm × 4 cm = 36 cm^2

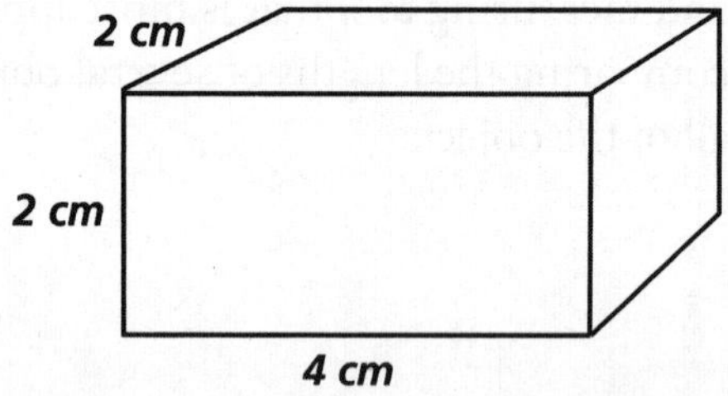

Volume = 4 cm × 2 cm × 2 cm = 16 cm^3

8 What are the length and width of the figure at the right?

Length ______________________

Width ______________________

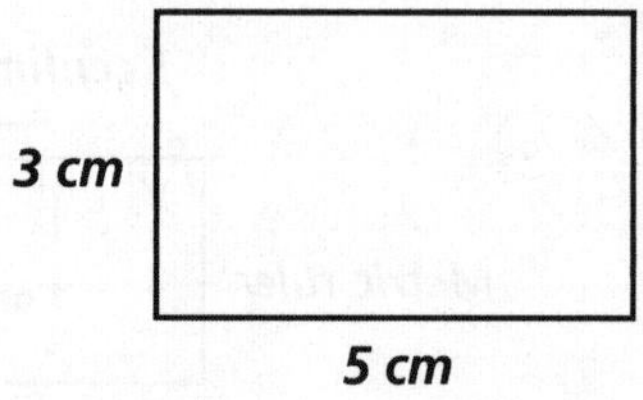

9 What is the area of the figure at the right? Be sure to use the correct units.

____

10 What are the length, width, and height of the figure shown on the right?

Length ______________________

Width ______________________

Height ______________________

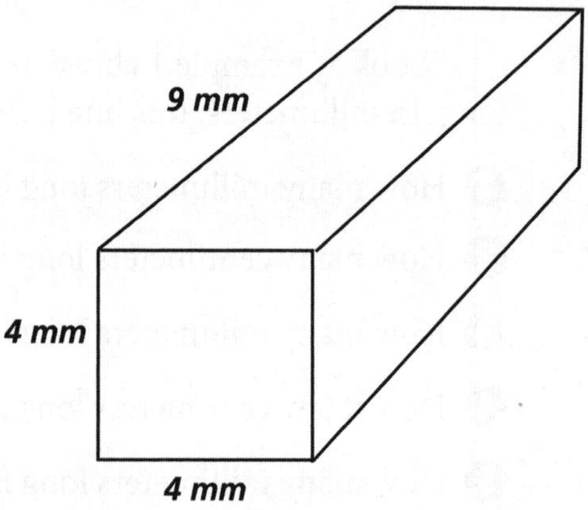

11 What is the volume of the figure on the right? Be sure to use the correct units.

____

12 **Think It Over** Find the area of a rectangle with a length of 5 cm and a width of 32 mm. Explain the steps you used to find your answer.

____

Copyright © Pearson Education, Inc., or its affiliates. All Rights Reserved.

Name ______________________ Date ______________ Class __________

INQUIRY SKILL FOCUS Practice

Measure: Liquid Volume

Write your answers to the questions below in the spaces provided. If you need more space, use a separate sheet of paper.

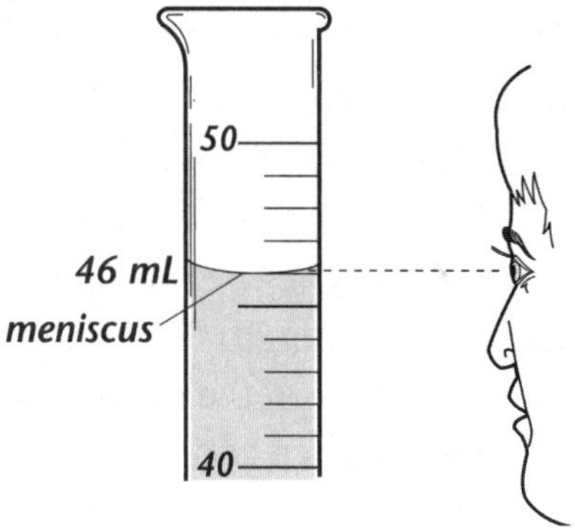

Volume is the amount of space an object takes up. A common SI unit for liquid volume is the liter (L). Liters are divided into smaller units, such as the milliliter (mL).

The graduated cylinder is a tool frequently used to measure liquid volume. ("Graduated" means that the cylinder is marked with measurement units.) Water in a graduated cylinder has a curved surface, as shown top right. The curved surface is called a meniscus. The volume should be read at the lowest point of the meniscus. Be sure to read the volume at eye level.

(*Hints:* Always check the unnumbered marks on a graduated cylinder to see what each mark indicates. Sometimes, the bottom of the meniscus falls between two markings. In that case, you need to estimate the measurement between the two marks.) Both of the graduated cylinders on the right contain the same volume of liquid. Although the cylinders have different markings, both cylinders contain 45 mL of liquid.

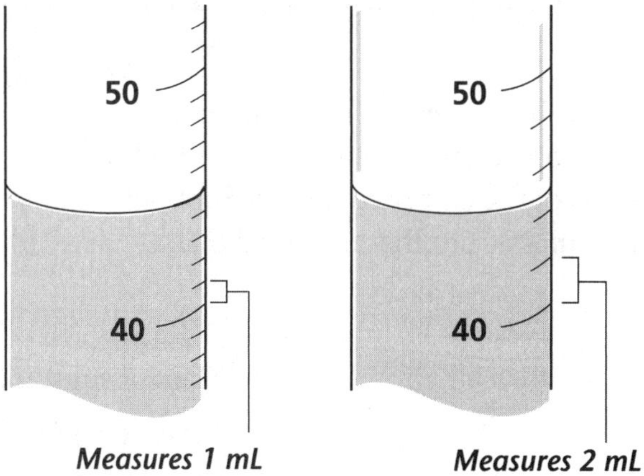

Measures 1 mL *Measures 2 mL*

What is the volume of liquid shown in graduated cylinders 1–4 below? What is the total volume in graduated cylinder 5?

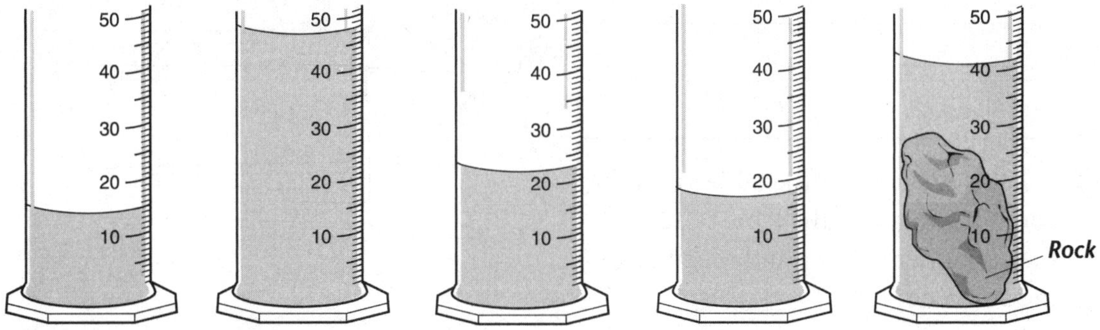

❶ ______ ❷ ______ ❸ ______ ❹ ______ ❺ ______

❻ If the diagrams for Questions 4 and 5 show the same graduated cylinder before and after the rock was added, what is the volume of the rock? ______________

❼ **Think It Over** Explain the steps that you would use to determine the value of the unnumbered marks on a graduated cylinder.

Copyright © Pearson Education, Inc., or its affiliates. All Rights Reserved.

Name ______________________ Date ____________ Class ________

INQUIRY SKILL FOCUS Practice

Measure: Mass

Write your answers to the questions below in the spaces provided.

Mass is the amount of matter in an object. Mass is measured using a balance. There are several different kinds of balances. Some balances give a single reading. Others give two or more readings that must be added to find the object's total mass.

The balance on the right is a triple-beam balance. The middle beam measures the largest amounts. To find the mass of an object, record the masses on each of the beams. Then add the readings.

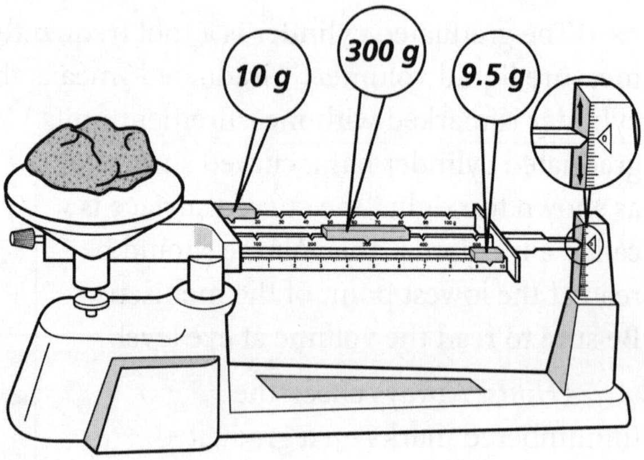

300 g + 10 g + 9.5 g = 319.5 g

(*Hint:* If you need to find the mass of an object in a container, first find the mass of the empty container. Subtract that mass from the combined mass.)

Mass of substance and container	246 g
Mass of container	−15 g
Mass of substance	231 g

1 The mass of the container holding this table salt is 3 g. What is the mass of the table salt?

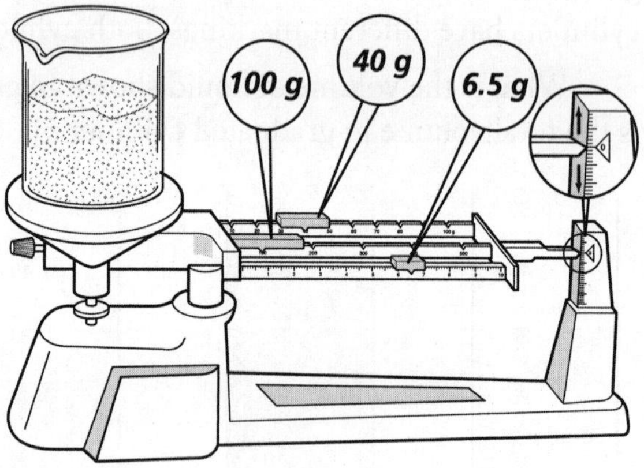

____

____

____

2 What is the mass of the table salt if the combined mass of the table salt and the container is 124 g and the mass of the container is 9 g?____________

____

3 Think It Over A gram is equal to 1,000 milligrams. If a sample has a mass of 52 grams, what is its mass in milligrams?

____

____

Copyright © Pearson Education, Inc., or its affiliates. All Rights Reserved.

Name ______________________ Date ____________ Class ________

INQUIRY SKILL FOCUS Practice

Measure: Temperature

Write your answers to the questions below in the space provided.

Temperature is a measure of how hot or cold something is. A Celsius thermometer, like the one shown here, is used to measure temperatures in science. The SI unit for temperature is °C, read as degrees Celsius. As you measure temperature using a thermometer, notice which thermometer marks are labeled and unlabeled. Check whether you are reading temperatures above or below zero. Temperatures below zero should be shown with a minus sign.

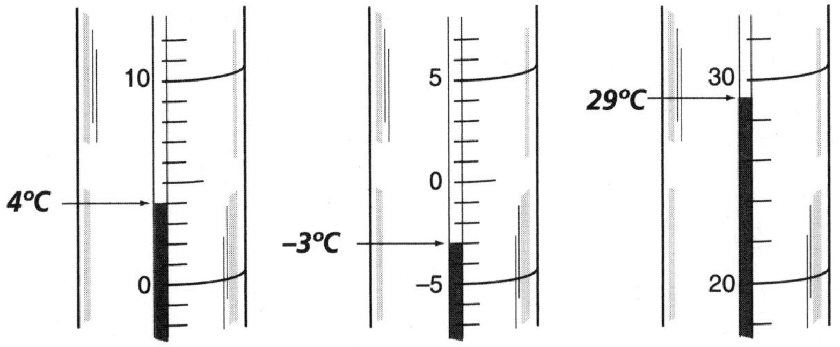

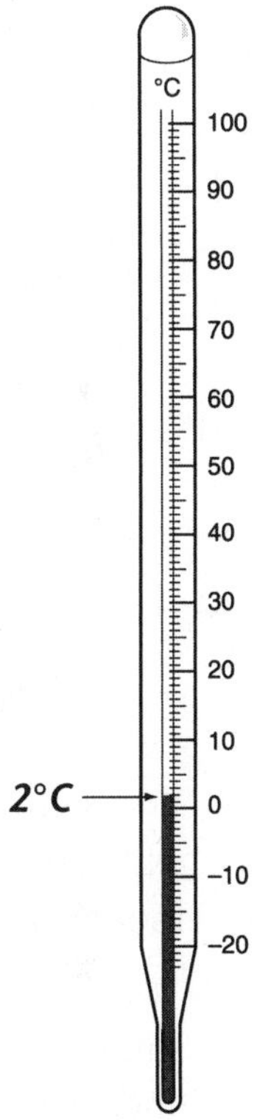

What temperature is shown in each of the diagrams below?

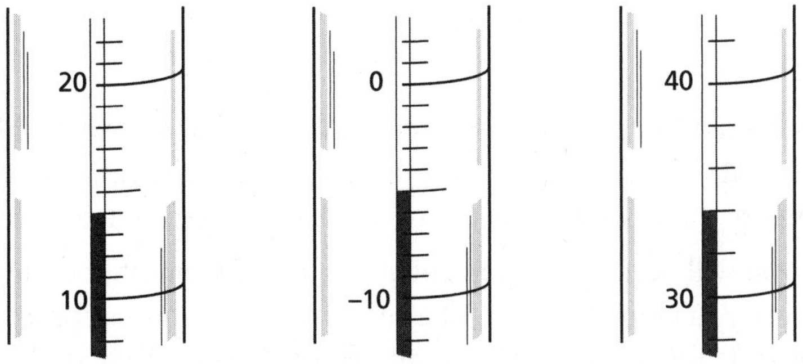

❶ ________ ❷ ________ ❸ ________

❹ The temperature of a water sample was 15°C. During an experiment, the temperature of the water increased by 16°C. What was the final temperature of the water? ______________________

❺ If a thermometer is numbered every 5 degrees and the temperature reading is exactly in the middle of the 25-degree and 30-degree markings, what is the temperature of the sample?______________________

❻ **Think It Over** If the outside temperature dropped from 7°C to –1°C, what was the total temperature change? Explain how you found your answer.

Copyright © Pearson Education, Inc., or its affiliates. All Rights Reserved.

Name ______________________________ Date ______________ Class __________

INQUIRY SKILL FOCUS Introduction

Calculate

Calculating is a process in which you use mathematical operations such as addition, subtraction, multiplication, and division to manipulate numbers and symbols. Calculating is a useful skill in everyday life as well as in many other areas, such as science, engineering, and technology.

One common kind of calculation is determining percentages. A percentage (%) is a number compared to 100. For example, 12% is $\frac{12}{100}$. Imagine a student collecting rocks for a school project. So far, she has collected 20 rocks. Of the 20 rocks, 5 are metamorphic. How could the student determine the percentage of metamorphic rocks in her rock collection?

First, write the comparison as a fraction.

$$5 \text{ out of } 20 = \frac{5}{20}$$

Next, multiply the fraction by 100%.

$$\frac{5}{20} \times \frac{100}{1} = \frac{500}{20}$$

Now, divide to find the percentage.

$$\frac{500}{20}\% = 25\%$$

Percentages are used in many areas of science. The efficiency of machines is expressed as a percentage. Food labels show the percentage of various nutrients in foods. The likelihood of rain or snow is often forecast as a percentage.

Sometimes you will need to use percentages to find the number or amount of something. Suppose you know that 15% of a certain food's Calories come from fat, and a serving contains 140 Calories. How could you determine the number of fat Calories in a serving?

The key to solving this problem is converting the percentage to a decimal. Any percentage can be converted to a decimal by first writing it as a fraction and then dividing by 100.

$$15\% = \frac{15}{100} = 0.15$$

Next, multiply the total number of Calories by the decimal form of the percentage.

$$15\% \text{ of } 140 = \frac{15}{10\cancel{0}} \times \frac{14\cancel{0}}{1} = \frac{^{3}\cancel{15}}{\cancel{2}\cancel{10}_{1}} \times \frac{\cancel{14}^{7}}{1} = 21 \text{ Calories from fat}$$

Copyright © Pearson Education, Inc., or its affiliates. All Rights Reserved.

Calculate *(continued)*

What if your science teacher tells you that 60% of the substances she tested dissolved in water? She found 18 substances that dissolve in water. What is the total number of substances she tested?

In this case we know that 18 is 60% of the total.

First, convert 60% to a decimal.

$$60\% = \frac{60}{100} = 0.60$$

Then, set up an equation.

$$x \times 0.60 = 18$$

$$x = \frac{18}{.60}$$

$$x = 30 \text{ total substances}$$

TIPS FOR CALCULATING

- If you perform calculations as a part of your research, show your work and each step of your calculation.
- Before you begin your calculation, be sure you have all the information you need.
- To convert a percentage to a decimal, divide the percentage by 100.
- Label your answer with the correct units.
- If your answer is a percentage, be sure to include the percent sign (%).
- Review your work. Have you correctly performed the calculations?

Checkpoint If you toss a coin 500 times, it's likely that 50% of the time the results will be heads and 50% of the time the results will be tails. How many times can you expect heads? How many times can you expect tails?

Copyright © Pearson Education, Inc., or its affiliates. All Rights Reserved.

Name ______________________________ Date ______________ Class __________

INQUIRY SKILL FOCUS Practice

Calculate

Write your answers on the lines below. Use the space provided to show your work.

1. Students recorded the weather each day for one month. In total, they had weather data for 30 days. On 6 of the days it was raining. On what percentage of days was it raining?

____

2. A black mouse and a brown mouse produced a litter of 8 offspring. Of the offspring, 6 were black and 2 were brown. What percentage of the offspring were black?

____

3. A serving of cereal has 22 Calories from fat. Twenty percent of the cereal's Calories come from fat. How many total Calories are in a serving of this cereal?

____

4. To use a machine, you must do work on the machine, called input work. The machine then does work on an object, called output work. Not all the input work gets transferred to output work. Some work is lost due to friction in the machine. If a machine has an input work of 250 joules and 25% of the input work is lost to friction, what is the output work of the machine?

____

5. **Think It Over** You can calculate some percentages in your head. Explain how you can calculate 50% of a number without using paper and pencil.

____

____

____

____

____

____

____

____

____

Copyright © Pearson Education, Inc., or its affiliates. All Rights Reserved.

Name ______________________ Date ____________ Class __________

INQUIRY SKILL FOCUS Introduction

Design an Experiment

In science, experiments are carefully planned in advance. When you make an organized plan to test an hypothesis, you are **designing an experiment**. Although every experiment is different, there are some steps that are important to all investigations. There are several different science skills used to plan investigations.

Pose a Question

In science, experiments are designed to answer questions or to solve problems. Suppose you have decided to plant a vegetable garden this summer. You know that your neighbor uses fertilizer in her vegetable garden. You wonder if you should use fertilizer in your garden as well. To find out, you could perform an experiment. The first step of the experiment is to ask a scientific question: Do vegetables grow larger with fertilizer than they do without fertilizer?

Develop a Hypothesis

The next step is to write a hypothesis, a possible explanation for a set of observations or a way to answer to a scientific question. A scientific hypothesis must be testable. Hypotheses are sometimes written as *If…then…*sentences. You might decide to design an experiment to test this hypothesis: *"If I use fertilizer, then the vegetables in my garden will grow larger."*

Plan the Procedure

After you've developed a hypothesis, you must plan the procedure you will use to test it. The procedure tells what tests you will perform. It also describes what data you will collect. A complete list of the needed materials is included in the procedure. The procedure also describes the variables in the experiment.

Control Variables In an experiment only one factor, or variable, should be changed at a time. The **manipulated variable** is the variable you deliberately change during an experiment. In the fertilizer experiment, the presence or absence of fertilizer is the manipulated variable. The **responding variable** is the variable that changes as a result of changes in the manipulated variable. The size of the vegetables is the responding variable in the fertilizer experiment. All of the other variables in the experiment must be controlled, or kept the same. The plants must receive the same amount of light and water, and be grown in the same soil. When the variables are controlled, you can be sure that changes in the rate of plant growth are due to the presence of fertilizer.

Copyright © Pearson Education, Inc., or its affiliates. All Rights Reserved.

Design an Experiment *(continued)*

Write Operational Definitions An operational definition is a statement that tells how a variable is measured. It allows others to repeat your experiment and understand your results. In the experiment with fertilizer, you would need to define how the size of the vegetables will be measured.

Interpret the Data

After the experiment is complete, you should review the data that were collected. Interpreting the data means explaining the data. You might look for trends, or patterns, in your results. Graphs, charts, and tables are useful for interpreting data.

Draw Conclusions

After the data have been interpreted, you can compare your results to your hypothesis. Then you can draw a conclusion using your results. A conclusion tells what was learned in the experiment. The conclusion tells whether or not the hypothesis was supported by the data. You might need to complete several trials of an experiment before you can make a conclusion.

TIPS FOR DESIGNING AN EXPERIMENT

- Scientific experiments begin with a testable question.
- Use the question to develop a hypothesis. Use the If...then...form to write a hypothesis.
- Make a plan that describes the procedure you will follow to test the hypothesis. Make sure you have all of the materials you will need to perform the experiment.
- Be certain you have identified the manipulated variable and the responding variable. Also identify the variables that need to be controlled.
- Write an operational definition for any terms that do not have a single, clear meaning.
- Collect the data carefully and accurately. Use data tables, charts, or graphs to show trends in the data.
- Use your data to draw conclusions.

Checkpoint Can an experiment have two manipulated variables? Why or why not?

____

____

____

Copyright © Pearson Education, Inc., or its affiliates. All Rights Reserved.

Name ______________________ Date ____________ Class __________

INQUIRY SKILL FOCUS Introduction

Pose Questions

Newspaper reporters often ask the "5 Ws"—Who? What? Where? When? and Why?—to gather information. Scientists also have a specific way of asking questions to gather information.

Scientific questions are questions about the natural world. Only questions that can be answered by observing, measuring, or investigating are considered scientific questions. Any question that is answered with an opinion or someone's personal values is not a scientific question. Look at these examples:

Question 1: Which type of apple tree grows most quickly? (scientific question)

Question 2: Which type of apple tastes best? (not a scientific question, because it is based on an opinion)

Some questions in science are very general. Although general questions are useful, they need to be made more specific before they can be answered in a single investigation. Scientists use specific questions to plan investigations.

Question 3: How does pollution affect an ecosystem? (general)

Question 4: Does the amount of water pollution affect the population of fish in a river? (specific)

Questions 5: How does the amount of water pollution affect the growth of trout? (more specific)

TIPS FOR POSING QUESTIONS

- Scientific questions are questions about the natural world that can be answered using observation, measurement, or controlled investigations.
- Check your questions to be sure they do not involve personal values or opinions.
- General questions can be used to think about general topics. To plan a scientific investigation, divide general questions into specific questions.

Checkpoint Write an example of a specific scientific question. Describe the investigation you could use to find the answer to this question.

____

____

____

____

____

Copyright © Pearson Education, Inc., or its affiliates. All Rights Reserved.

Name ______________________ Date ____________ Class __________

INQUIRY SKILL FOCUS Practice

Pose Questions

Examine the statements below. For each of Questions 1–5, write yes if the topic can be investigated scientifically. Write no if it cannot be investigated scientifically. Then, for each item to which you answered yes, rewrite the statement in the form of a scientific question. Write your answers in the space provided. If you need more space, use a separate sheet of paper.

1. The more quickly water flows, the more erosion it causes.

2. It is the responsibility of farmers to find ways to reduce soil erosion.

3. Soil is one of the most important natural resources in the world.

4. The composition of soil affects the growth rate of plants.

5. Some rocks break down into soil more quickly than others.

For each of the general questions below, develop two specific questions that could be answered in an investigation.

6. How do living things in an area interact with one another?

7. What are some ways that plant species differ from one another?

8. **Think It Over** Why are questions about opinion or personal values considered unscientific?

Copyright © Pearson Education, Inc., or its affiliates. All Rights Reserved.

Name ______________________________ Date ____________ Class __________

INQUIRY SKILL FOCUS Introduction

Develop a Hypothesis

Imagine that you are mixing sugar with icewater and lemon juice to make homemade lemonade. You stir and stir to get the sugar to dissolve. At the same time, your mother makes a cup of hot tea and easily gets the sugar to dissolve. Why are you having so much trouble getting the sugar to dissolve? What's causing the difference between the lemonade and the tea?

Questions, like this one about the lemonade, can lead to hypotheses. A hypothesis (plural: hypotheses) is a possible explanation for a set of observations or answer to a scientific question. Hypotheses must be testable. In other words, you should be able to find out if a hypothesis is true or not by performing an investigation or making observations. Writing your hypothesis in the form of an If…then… statement lets you be sure the hypothesis is testable.

These examples show how scientific questions can be used to write properly worded hypotheses.

Scientific Question: Does the lemon juice in the lemonade keep the sugar from dissolving quickly? (a specific scientific question that does not involve personal opinion)

Hypothesis: If a drink contains lemon juice, then sugar will dissolve more slowly than in other drinks. (a testable and properly worded hypothesis)

Scientific Question: Does the temperature of the drink affect how quickly the sugar dissolves? (a specific scientific question that does not involve personal opinion)

Hypothesis: If a drink is warmer, then sugar will dissolve in it more rapidly than it does in a cooler drink. (a testable and properly worded hypothesis)

TIPS FOR DEVELOPING HYPOTHESES

- Scientific questions are used to make hypotheses. A scientific question is a testable question about the natural world.
- Use the *If…then…*form to write a hypothesis.
- Review your hypothesis to be sure it can be tested through an investigation.
- Remember that hypotheses are not necessarily true. An investigation can prove a hypothesis to be incorrect.

Checkpoint Write a scientific question and a hypothesis about seed germination.

____

Copyright © Pearson Education, Inc., or its affiliates. All Rights Reserved.

INQUIRY SKILL FOCUS Practice

Develop a Hypothesis

Use the diagrams to answer the questions on the next page.

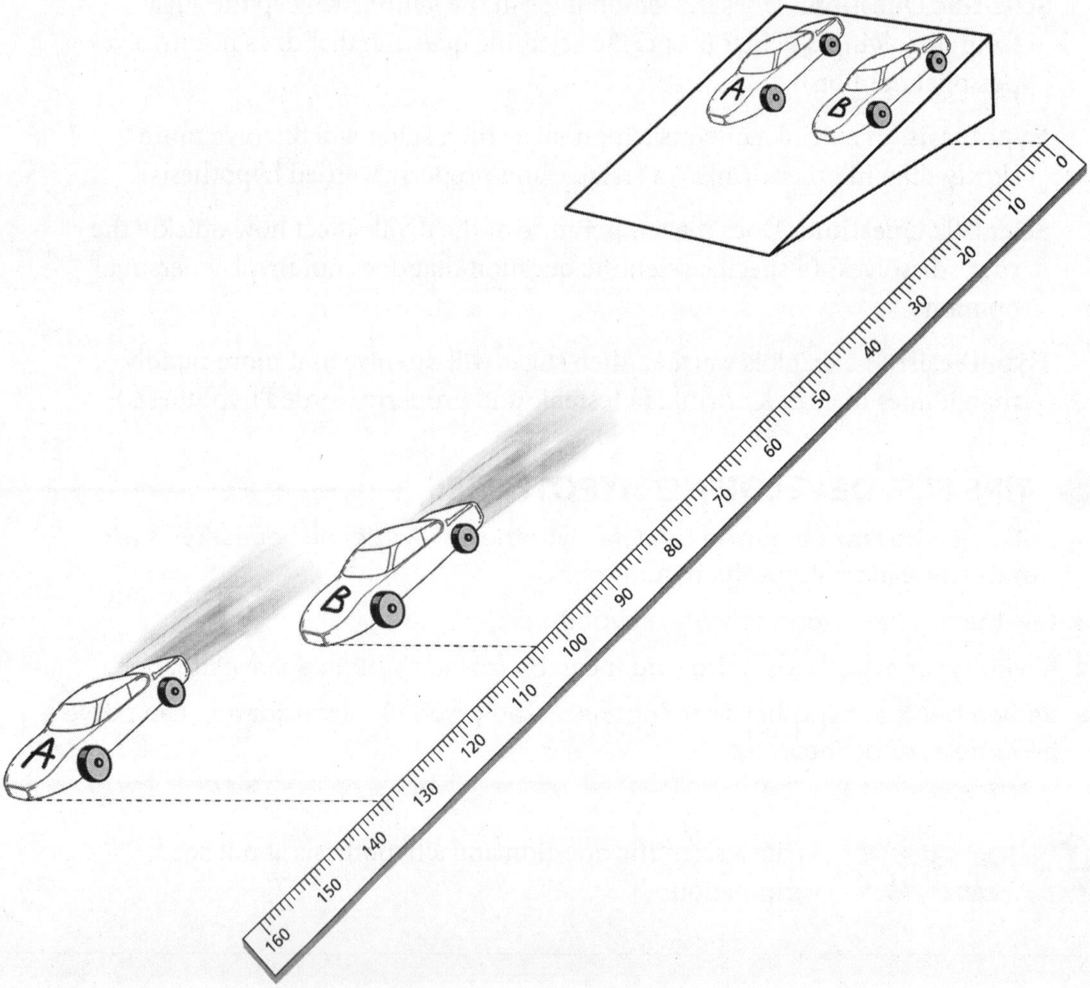

Copyright © Pearson Education, Inc., or its affiliates. All Rights Reserved.

Name ______________________________ Date ____________ Class ________

Develop a Hypothesis *(continued)*

Answer the following questions. Use a separate sheet of paper if you need more space.

1. Look at the figure on the previous page. Observe the motion of cars A and B. Use your observations to write a specific scientific question about the motion of cars A and B.

____

____

2. Use your question to develop a hypothesis that could be tested. Use the *If… then…* form to write your hypothesis.

____

____

3. How could your hypothesis be tested? Write a brief description of an investigation that could test your hypothesis. Be sure to explain what variables will affect the cars' motion.

____

____

____

4. Review your hypothesis. List two other possible explanations that could describe any differences in the cars' motion.

____

____

____

5. Write two more hypotheses in the If…then… form based on the two possible explanations you developed to answer Question 4. How can you find out which of your three hypotheses is most correct?

____

____

____

6. **Think It Over** Is a hypothesis always a correct, factual statement?

____

____

____

Copyright © Pearson Education, Inc., or its affiliates. All Rights Reserved.

Name ______________________________ Date ______________ Class __________

INQUIRY SKILL FOCUS Introduction

Control Variables

Imagine trying a recipe for homemade granola bars. When you tasted them, you weren't pleased with the results. The next time you made granola bars, you used more oatmeal, fewer raisins, a lower oven temperature, and a longer baking time. This time, the granola bars tasted great! What made the difference between the two batches? Since more than one factor was changed, it is impossible to know.

Variables are the factors in an experiment that can change. In the example above, the amount of oatmeal, the number of raisins, the oven temperature, and the baking time are all variables. In a scientific experiment, only one variable is changed at a time. The variable that is purposely changed is called the **manipulated variable.** The factor that might change as a result of the manipulated variable is the **responding variable.**

In a scientific experiment, all of the other variables must be controlled. **Controlling variables** means keeping all conditions the same except for the manipulated variable. Imagine an experiment to test how air temperature affects the rate at which ice melts.

Manipulated Variable: air temperature

Responding Variable: rate at which the ice melts

Controlled Variables: amount of ice; shape of ice (cube, shaved, block); type of container

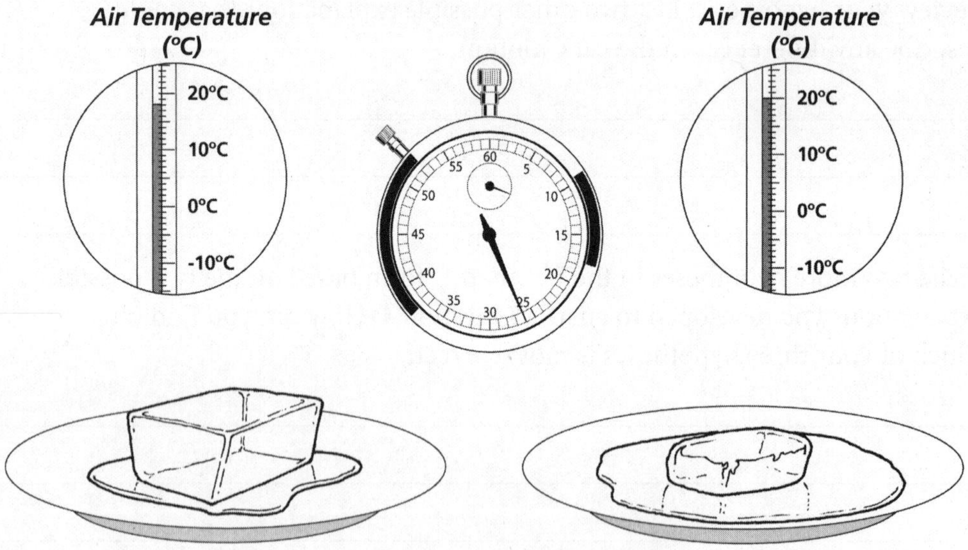

In a scientific experiment, only the manipulated variable is changed. All of the other variables are controlled. You can be sure that changes in the responding variable are due to changes in the manipulated variable. In this particular experiment, you can tell that changes in the rate at which the ice melts are due to differences in temperature. When all of the variables except the manipulated variable are controlled, the experiment is called a **controlled experiment.**

Copyright © Pearson Education, Inc., or its affiliates. All Rights Reserved.

Name ______________________ Date ______________ Class __________

Control Variables *(continued)*

Experimental Groups and Control Groups Suppose a scientist wants to study how fertilizer affects the growth rate of tomato plants. To carry out an experiment, the scientist uses a set of plants that is divided into two groups. The first group, or experimental group, consists of tomato plants that are given fertilizer. The second group consists of plants that are *not* given fertilizer. This group, known as the control group, would then show how fast such plants would grow without any fertilizer.

The purpose of the control group in an experiment is to serve as a standard of comparison. Both the control group and the experimental group are tested under the same conditions. In other words, all the plants are about the same size and type and in the same kind of soil, and they are given the same amount of water and sunlight. The scientist can then compare the experimental group to the control group to see how the fertilizer affected the rate of growth.

TIPS FOR CONTROLLING VARIABLES

- Use your hypothesis to plan your experiment.
- Identify the manipulated variable and the responding variable in the experiment. Make a plan for how you will change the manipulated variable. Decide how you will measure changes in the responding variable.
- Make a list of all of the other variables that could affect the responding variable. Make a plan to control these variables.
- Decide if your experiment calls for a control group. If so, plan which group of objects or living things will serve as the control group.

Checkpoint Make a list of at least three variables that would need to be controlled in an experiment to determine the effect of different-colored lights on plant growth.

____

____

____

____

____

____

____

Copyright © Pearson Education, Inc., or its affiliates. All Rights Reserved.

Name ______________________ Date ____________ Class ________

INQUIRY SKILL FOCUS Practice

Control Variables

Answer the following questions in the space provided. Use a separate sheet of paper if you need more space.

1. In an experiment to determine the effect of different temperatures on the rate at which bananas ripen, what is the manipulated variable and the responding variable?

2. Name the manipulated variable and the responding variable in an experiment to determine if the mass of a toy car affects the rate at which it rolls.

3. Name three variables that would need to be controlled in the experiment described in Question 2.

4. Describe the control group and the experimental group in an experiment to determine whether using antibacterial soap affects the total number of illnesses a person gets in one month.

5. Why would it be important to use groups, rather than individuals, in the experiment described in Question 4?

6. **Think It Over** In a student's lab report, she listed two manipulated variables. Could this student make valid conclusions from the results of her experiment? Why or why not?

Copyright © Pearson Education, Inc., or its affiliates. All Rights Reserved.

INQUIRY SKILL FOCUS Introduction

Form Operational Definitions

Suppose the school librarian asked you and one of your friends to organize the books on the library shelves. You decide to organize the books by size. Your friend chooses to organize the books by topic. Unfortunately, the librarian wanted you to organize the books by author. Since you did not have a clear definition of the term *organize the books,* each person chose a different method.

In science, exact definitions are important. Scientists often run several trials of their experiments. They also repeat the work of other scientists. To do so, they need terms that are clearly defined. An **operational definition** is a statement that describes how a particular variable is to be measured, or how an object or condition is to be recognized. (The word *operational* means "describing what to do.") Operational definitions are used whenever a term does not have a single, clear meaning.

Imagine an experiment in which you measure the effect of a healthy diet on the activity of hamsters. The term *healthy diet* does not have a single, clear meaning. It must be defined. How will the *activity* of the hamsters be measured? An operational definition will need to be provided to explain how this variable will be measured.

In the experiment described above, the term *healthy diet* might have the following operational definition: 25 grams per day of Healthy Hamster brand food was provided for each hamster. An operational definition describing how to measure activity might be: The hamster's activities were recorded every hour. A hamster engaged in any activity other than sleeping was considered active.

When you write an operational definition, make sure it describes what to do or what to observe. Be sure that any terms in your lab report that do not have a single, clear meaning have an operational definition. Here are some examples of operational definitions.

Notice how this operational definition defines a term that does not have a single, clear meaning:

Example 1: The plants were grown in a sunny area. Operational definition: A sunny area is an area that receives an average of 10 or more hours of sunlight each day.

This operational definition tells how a variable will be measured:

Example 2: Toy cars roll faster if they have larger wheels. Operational definition: The size of the wheels was determined using a centimeter ruler to measure the diameter of the wheels.

Copyright © Pearson Education, Inc., or its affiliates. All Rights Reserved.

Name ____________________ Date __________ Class __________

Form Operational Definitions *(continued)*

The same term can have a different operational definition in different experiments. The term *growth of the plant* can be defined in many ways. Here are some examples: (1) the height of the plant from the soil surface to the plant's highest part; (2) the mass of the total plant, including the roots, measured when dry; (3) the number of leaves that have sprouted. The terms *large plant* and *small plant* might have very different meanings in different investigations.

Some variables can be measured in a variety of ways. You need to choose a definition that works well for a particular experiment. A good operational definition for the measurement of a variable includes the tool used to measure and the unit of measure used. It should also tell how frequently a particular measurement is made. In an experiment to compare plant growth rates, you might use the following operational definition: The plant's growth will be measured every morning using a centimeter ruler. The total height of the plant will be rounded to the nearest millimeter and recorded.

TIPS FOR WRITING OPERATIONAL DEFINITIONS

- Make a written plan for your experiment or investigation.
- Identify any terms that require an operational definition. Write a clear, complete definition for these terms.
- If there is more than one possible way a variable could be measured, write an operational definition that provides an exact description of the method you will use.
- Read through your definitions. Could you use the information to repeat your experiment exactly?
- Have a friend review your written plan. Ask if your friend could duplicate your work using only the written plan. If not, revise your operational definitions.

Checkpoint Explain why there can be more than one correct operational definition for the term *cold.*

____

____

____

____

____

____

____

Copyright © Pearson Education, Inc., or its affiliates. All Rights Reserved.

Name ______________________ Date ____________ Class __________

INQUIRY SKILL FOCUS Practice

Form Operational Definitions

Write an operational definition for each underlined idea in the space provided. Use a separate sheet of paper if you need more space.

1. The plant grew quickly.

____

____

____

2. Investigate two brands of paper towels to see which is stronger.

____

____

____

3. In the autumn, trees lose their leaves more quickly on windy days.

____

____

____

4. Measure the temperature of a beaker of water.

____

____

____

5. Certain breeds of dogs have long legs.

____

____

____

6. **Think It Over** Look at the definition you wrote for Question 5. How would your operational definition for the term *long legs* be different if you were studying horses? Explain why some terms can have more than one correct operational definition.

____

____

____

Copyright © Pearson Education, Inc., or its affiliates. All Rights Reserved.

Name ______________________ Date ____________ Class ________

INQUIRY SKILL FOCUS Introduction

Interpret Data

Imagine that you are in charge of an aquarium that will be kept in your science classroom. How will you decide what types of fish you will keep? To make a good decision, you must know what size aquarium you will use. You need to know what types of fish can live together. You need to know how much space and what water conditions each type of fish requires. Using all this information, you can make a plan for the classroom aquarium. Carefully examining data to make decisions is similar to the skill of interpreting data, an important part of scientific investigations.

The observations and measurements that are made during a scientific investigation are called **data.** After all of the data for an investigation have been collected, they need to be carefully examined. **Interpreting data** is the process of finding meaning in data by looking for patterns and trends.

Suppose you collected data about the amount of a particular substance that would dissolve in different volumes of water. You and your lab partner made the following data table with your results.

Amount Dissolved in Different Volumes of Water

Volume of Water (mL)	Amount Dissolved (g)
50	3
100	6
150	9
200	12
250	15

When you and your lab partner examine the data table, you can see that the mass of the substance that dissolves increases as the volume of water increases. To further examine the trend in the data, you make a graph.

Copyright © Pearson Education, Inc., or its affiliates. All Rights Reserved.

Name ______________________ Date ____________ Class __________

Interpret Data *(continued)*

This line graph would help you to interpret the data from your experiment. Here are some of the inferences you could make using the graphs.

Example 1: A greater mass of the substance will dissolve in a greater volume of water.

Example 2: For every 50 mL increase in the amount of water, the mass of the substance dissolved increases by 3 g.

Example 3: The mass of this substance dissolved in 300 mL of water would be 18 g.

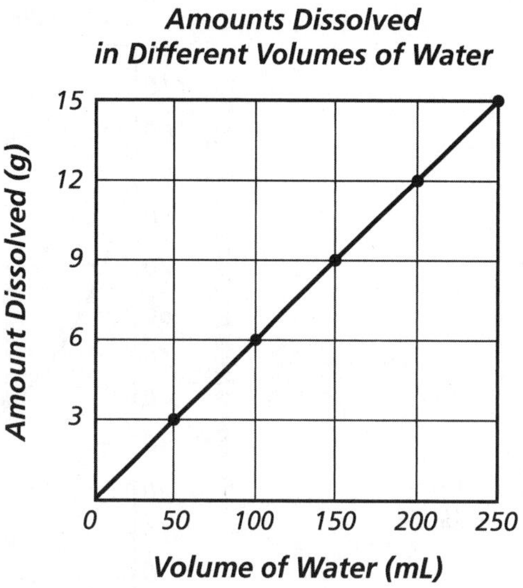

To determine if your inferences are logical, think about what you have learned about solutions in science class. You know that the amount of a substance that will dissolve depends on several factors. You know that a certain amount of a substance will dissolve in a certain volume of water. It makes sense to you that the greater the volume of water, the greater the mass of substance that you can dissolve in it. You and your lab partner decide that your inferences are logical.

TIPS FOR INTERPRETING DATA

- During investigations, use a data table or chart to collect and organize your data.
- Whenever possible, make a graph using the data.
- Identify trends, or patterns, in the data.
- Use the data to make inferences. Do your inferences make sense compared with what you already know about a topic? If not, review your work.

Checkpoint Describe two reasons that your inferences about a set of data might not match your previous knowledge of a topic.

____

____

____

____

____

Copyright © Pearson Education, Inc., or its affiliates. All Rights Reserved.

Name ______________________ Date ______________ Class ________

INQUIRY SKILL FOCUS Practice

Interpret Data

Answer the following questions on a separate sheet of paper.

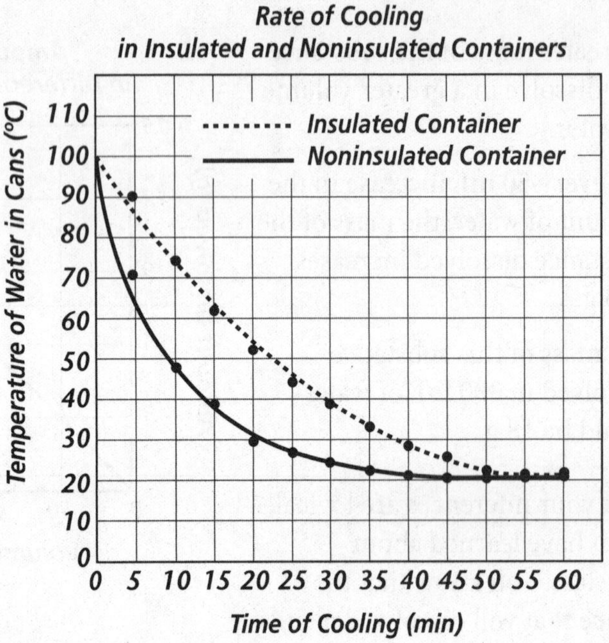

1. Write a sentence that summarizes the data. Use the title of the graph and the axis labels to help you write your sentence.
2. Name three times at which the water in the two containers have the same temperature.
3. Name one time at which the temperatures of the water in the two cans differed by more than 20°C.
4. Compare the rates at which the water in the two containers cooled.
5. What do you expect the temperatures of the water in the containers would be at 65 minutes? Explain.
6. **Think It Over** Substances that are insulators of heat slow changes in temperature. Knowing this fact, is your answer to Question 4 logical? Why or why not?

____

____

____

____

Copyright © Pearson Education, Inc., or its affiliates. All Rights Reserved.

Name ______________________________ Date ______________ Class ____________

INQUIRY SKILL FOCUS Introduction

Draw Conclusions

Suppose you find that your basketball is flat. You pump air into the basketball, but it goes flat again. You would probably conclude that there is a leak in the ball. We all make conclusions in everyday situations. In science, the word *conclusion* has a more limited meaning. **Drawing a conclusion** means making a statement summing up what you have learned from an experiment.

Recall that experiments begin with a hypothesis often written in the *If…then…*form. After the experiment is complete, the data are examined to see whether the hypothesis is correct or incorrect. Using the hypothesis and the data, you can write a conclusion that tells what you learned from the experiment. Scientists may repeat their work and compare their work to the work of others before they draw their final conclusions.

Imagine that you and your lab partner have decided to experiment to see if the strength of Brand Y paper towels is changed when they get wet. Your lab partner saw an advertisement that convinced her that Brand Y paper towels have the same strength no matter if they are wet and dry.

This is her hypothesis: *If a Brand Y paper towel is wet, then it will be just as strong as when it was dry.*

You don't think the advertisement is true. Here is your hypothesis: *If a Brand Y paper towel is wet, then it is weaker than a dry Brand Y paper towel.*

The following data were collected during the experiment.

Brand Y Paper Towel Strength Investigation

Dry Conditions		Wet Conditions	
Trial	Mass Held Before Tearing (g)	Trial	Mass Held Before Tearing (g)
1	86	1	57
2	89	2	51
3	86	3	57
4	88	4	52
5	86	5	53
Average	87	Average	54

After the controlled experiment is complete, you and your lab partner examine the data. Your data clearly show that wet Brand Y paper towels are not as strong as dry Brand Y paper towels.

Copyright © Pearson Education, Inc., or its affiliates. All Rights Reserved.

Name ______________________ Date ____________ Class ________

Draw Conclusions *(continued)*

Here is the conclusion your lab partner writes: *Based on the results of a controlled experiment, wet Brand Y paper towels are not as strong as dry Brand Y paper towels.*

Your conclusion might be: *Wet Brand Y paper towels are weaker than dry Brand Y paper towels.*

It is important to draw conclusions that are based only on the experimental data. Imagine that you drew the following conclusion from the experiment: *Paper towels are weaker when they are wet than when they are dry.*

You cannot reasonably draw this conclusion based on the data from your experiment, because you did not test all brands of paper towels. Only conclusions about Brand Y paper towels can be drawn using your data.

TIPS FOR DRAWING CONCLUSIONS

- Review the hypothesis for the experiment.
- Examine the observations and measurements you made during the experiment. Analyze your data using graphs, tables, and charts.
- Decide whether you need to repeat the experiment. If possible, compare your work to the work of others.
- Write a conclusion that sums up the results of the experiment.
- Review your conclusion. Be certain that have used only your experimental data to make your conclusion.

Checkpoint Can you make a conclusion if your data do not agree with your hypothesis? Why or why not?

Copyright © Pearson Education, Inc., or its affiliates. All Rights Reserved.

Name ______________________ Date ____________ Class ________

INQUIRY SKILL FOCUS Practice

Draw Conclusions

Answer the following questions on a separate sheet of paper.

You and a friend have decided to experiment to find the effects of chlorinated tap water on the growth rate of the aquarium plant elodea. You make the following hypothesis: *If there is chlorine in the elodea's water, then it will grow more slowly.* You've planned a controlled experiment, which you carry out over the course of several weeks. To analyze your data, you make the following graph.

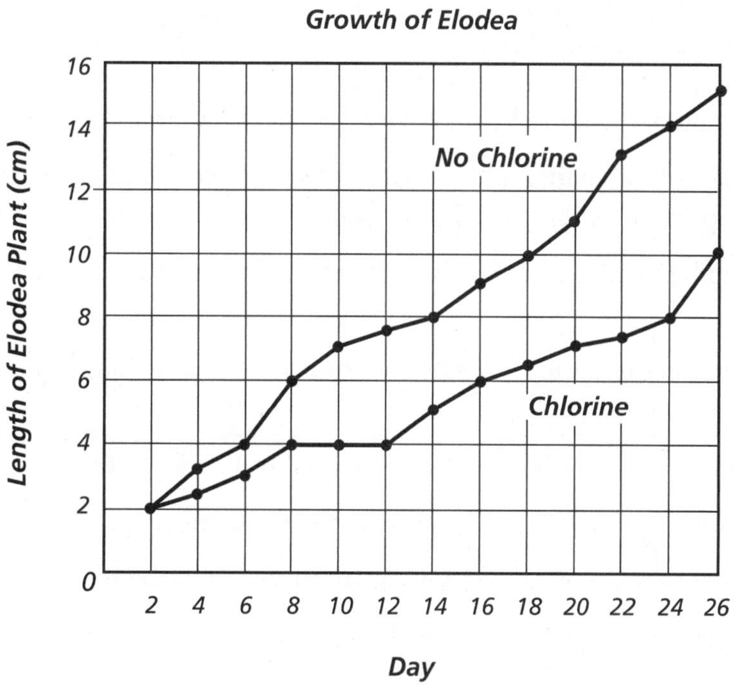

1. Write a sentence that summarizes the data in the graph.
2. Do the data show that your hypothesis is correct or incorrect?
3. Write a conclusion for this experiment.
4. Explain whether you can reasonably draw this conclusion from your data: *Aquarium plants grow more slowly if there is chlorine in their water.*
5. **Think It Over** Imagine you made the following hypothesis: *If there is chlorine in the elodea's water, then the plant will grow more quickly.* Would you have learned anything from the experiment? Why or why not?

____

____

____

Copyright © Pearson Education, Inc., or its affiliates. All Rights Reserved.

Name ______________________________ Date ______________ Class __________

INQUIRY SKILL FOCUS Practice

Design an Experiment

Choose a question from the list below as a topic for an experiment. Alternatively, pose a scientific question of your own and obtain your teacher's approval to use that question. Remember, as one of the first steps in planning your investigation, you may need to narrow your original question. Then write a hypothesis and design an experiment to answer the question. Be sure to include all the necessary parts of an experiment, such as naming the manipulated and responding variables and identifying the variables you will control. Write any operational definitions that are needed. Include a data table you could use for recording your observations. Use a separate sheet of paper for your work.

1. Does one brand of plant food work better than another brand?
2. Does the color of a container affect the rate at which water in the container changes temperature when placed in sunlight?
3. Does the depth at which bean seeds are planted affect the rate at which they sprout?
4. Does the amount of salt in a saltwater solution affect the boiling point of the solution?
5. Does the mass of a baseball affect the distance it travels when hit?
6. Does the presence of fish affect the growth rate of aquarium plants?
7. Does friction affect the distance a toy car travels when pushed?
8. How is the growth of bacteria affected by antibacterial dish soap?
9. How does the particle size of soil affect the rate at which water flows through it?
10. Does the amount of stirring affect the rate at which salt dissolves in water?
11. **Think It Over** Why is it important to include operational definitions in your experimental design?

____

____

____

____

____

____

____

Copyright © Pearson Education, Inc., or its affiliates. All Rights Reserved.

Name ________________ Date ________ Class ________

INQUIRY SKILL FOCUS Introduction

Create Data Tables

Imagine that you and your friends have a lawn-mowing business. You need to keep track of all your customers, their addresses, and their payment schedules. You could use a chart or table to organize all of this information.

During scientific investigations, there is often a large amount of information to organize. A **data table** is an organized arrangement of information in labeled rows and columns. Data tables are used during experiments to organize measurements and observations. After the experiment is complete, the data table can be used for data analysis.

Think about the manipulated variable and the responding variable when you plan your data table. Use this information to plan the columns and rows of the data table. How will these variables be observed or measured? What units of measure will be used? The units should be inserted in the column heads. Consider how frequently changes in the variables will be measured. Each data table should have a title.

Look at the data table below. This data table was made for an experiment to determine if air temperature affects the rate of plant growth. Notice that this data table includes space for all the required measurements, the units that will be used, and a descriptive title.

Effects of Air Temperature on Plant Growth

	Height of Plant (cm)		
Time (days)	15°C	25°C	35°C
2			
4			
6			

As you plan your experiment, make a rough draft of your data table. Compare the data table to the experimental design to be sure there is a space for each required measurement. Think about the units of measure in the column heads. Do they match the units of measure in the plan for the experiment? Some experiments require several trials. Be certain that there is space in your data table if several trials are needed. Also be sure there is space for observations or measurements of the control group. This will allow direct comparison of the experimental group with the control group when it is time to analyze the data.

Copyright © Pearson Education, Inc., or its affiliates. All Rights Reserved.

Name ______________________________ Date ______________ Class __________

Create Data Tables *(continued)*

The data table below shows one way of recording data from multiple trials of an experiment. Note that there is also space to record the average of the data from the three trials.

Effect of Ramp Angle on Distance Traveled by a Toy Car				
	Distance (cm)			
Angle of Ramp	**Trial 1**	**Trial 2**	**Trial 3**	**Average**
20°				
40°				
60°				

After a careful review of your rough draft, create your final data table in your notebook. Having the data table prepared in advance will allow you to focus all of your attention on the experiment as it progresses. It will also remind you to obtain all of the needed measurements and observations.

TIPS FOR CREATING DATA TABLES

- Identify the manipulated and the responding variable in the experiment to determine what observations you will make.
- If you need to make observations at a regular time interval, make sure your data table shows this. Include space for each of the required observations.
- Make a draft of the data table. Include labels on the rows and columns and a title for the table. Include the units of measure in the column labels.
- Make sure your draft has a place for all of the measurements required in your experiment. Be sure there is a place for observations of the control group. If your experiment involves multiple trials, be sure you have included these.
- Revise the draft of the data table as needed. Record the final version of your data table in your notebook.

Checkpoint Why is it important to have the data table planned and written in your notebook before an experiment begins?

____

____

____

____

Copyright © Pearson Education, Inc., or its affiliates. All Rights Reserved.

Name ______________________ Date ____________ Class __________

INQUIRY SKILL FOCUS Practice

Create Data Tables

Read the description of the experiment. Then answer the questions that follow on a separate sheet of paper.

We experimented to find some of the factors necessary for the growth of bread mold. We placed two identical slices of bread on separate plates. We moistened one of the slices of bread with tap water. The other slice was kept dry. Both slices were left out in the air for 1 hour. Then we put each slice of bread in a sealable bag. We put the bread in the bags in a warm, dark place. Each day for 5 days we measured in centimeters the area of mold on each slice of bread. We compared the growth of mold on the wet bread to the growth of mold on the dry bread.

1. Identify the manipulated variable and the responding variable in this experiment.
2. How many times will the responding variable be measured?
3. What units of measure will be used in this experiment?
4. Using the description of the experiment and the answers to your questions, create a complete data table for the experiment. Be sure that you have included units for each column and a title for the data table.

Copyright © Pearson Education, Inc., or its affiliates. All Rights Reserved.

Name ______________________ Date ______________ Class __________

INQUIRY SKILL FOCUS Practice

Create Data Tables

Read the description of the experiment. Then answer the questions on a separate sheet of paper.

We experimented to see if the surface area of salt affected how quickly it dissolved in water. We used rock salt and table salt. Each type of salt was dissolved in water with the same amount of stirring. We measured the time, in seconds, that it took for identical masses of the two different kinds of salt to dissolve. For each type of salt we performed three trials, and we averaged the results of the three trials.

5. How many trials does this experiment require?
6. What calculation using the experimental data is required?
7. How frequently will measurements be taken in this experiment?
8. Using the description of the experiment and the answers to your questions, make a complete data table for the experiment. Be sure you have included space for multiple trials. Also include space for the results of any required calculations.
9. **Think It Over** Look at the data table you made. Why is it important to include units in the column heads?

____

____

____

____

____

Copyright © Pearson Education, Inc., or its affiliates. All Rights Reserved.

Name ______________________________ Date ____________ Class ________

INQUIRY SKILL FOCUS Introduction

Create Bar Graphs

Your school has several choices of after-school clubs. Your principal has asked you to make a graph showing the number of members in each club. How could you use the following data to make a graph?

After-School Clubs	
Club	**Number of Members**
Computer Club	60
Poetry Club	30
Running Club	50
Science Club	25

The first step in graphing this data is choosing the appropriate type of graph. Your data tell the number of students in each club. These are separate categories. Bar graphs show data about separate but related categories. The bar graph shown below shows the number of students in each club.

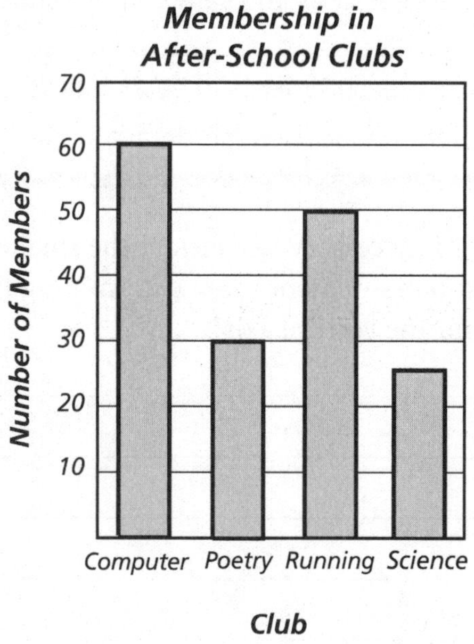

Each club has a bar that represents the number of members. Each club is listed on the horizontal axis. This axis also has a label that describes the categories. The vertical axis is marked with a scale and a label. In this case, each square on the graph paper represents 5 students. The vertical axis also has a label. The height of the bars indicates the number of students in each club. The title of the graph= describes the data.

Copyright © Pearson Education, Inc., or its affiliates. All Rights Reserved.

Create Bar Graphs *(continued)*

When you make a bar graph, it is important to choose a scale for the vertical axis that will accommodate all of the different measurements. Consider the largest and the smallest amount you will need to represent on your graph. Decide the amount that will be represented by each square on the graph paper.

Remember that both axes should be labeled. Include units of measure where appropriate. The individual bars also need labels. Be sure to give the graph a descriptive title.

TIPS FOR CONSTRUCTING BAR GRAPHS

- Use a data table to organize the data.
- On your graph paper, draw a vertical and horizontal axis for the graph.
- Determine the scale for the measurements shown on the vertical axis. The scale should be chosen by considering the largest and the smallest measurements. Each square on the graph paper will represent a certain amount.
- Label the vertical axis with the scale and the units of measure.
- Label the horizontal axis. Make a bar for each category on the horizontal axis. The width of all of the bars should be the same. There should be a space between each bar.
- Draw bars to represent the measurements in each category.
- Write a title for the graph that describes the data it shows.

Checkpoint Why is it necessary to consider the smallest and the largest measurement when determining the scale to be used for the vertical axis?

____

____

____

____

____

____

____

____

____

____

Copyright © Pearson Education, Inc., or its affiliates. All Rights Reserved.

Name ______________________ Date ____________ Class ________

INQUIRY SKILL FOCUS Practice

Create Bar Graphs

Answer the questions below on a separate sheet of paper. Use a sheet of graph paper to make the graph.

Sound waves travel at different speeds through different substances. The speed of sound in different materials is shown in the data table below. Use the data and the answers to your questions to make a bar graph representing the data.

Speed of Sound in Different Substances

Substance	Speed (m/s)
Plastic	1,800
Copper	3,100
Wood	4,000
Iron	5,100
Steel	5,200

1. What label will you use for the vertical axis?
2. Consider the smallest and the largest measurement you will need to represent on the graph. What scale will you use in your graph?
3. On a sheet of graph paper, make a bar graph that shows the data. Be sure to include a title for the graph and labels for each axis.
4. **Think It Over** What things should you consider when you plan a title for your graph?

Copyright © Pearson Education, Inc., or its affiliates. All Rights Reserved.

Name ____________________ Date ______________ Class __________

INQUIRY SKILL FOCUS Introduction

Create Line Graphs

Members of a science class were studying the effects of exercise on pulse rates. They measured the pulse rate of one student at regular time intervals as she warmed up, exercised, and cooled down. They used a data table to record their results.

Pulse Rate During Warm-Up, Exercise, and Cool-Down

Time (min)	Pulse Rate (beats per minute)
2	75
4	82
6	100
8	105
10	95
12	80

To show trends in the data more clearly, the class decided to make a line graph. A **line graph** is used to display data that show how one variable (the responding variable) changes in response to another variable (the manipulated variable). A line graph is used when the manipulated variable is continuous, that is, when there are other measurements possible between the measurements you recorded. In this example, time is a continuous variable. The students could have measured the student's pulse rate every minute, for example. The line graph of the data is shown below.

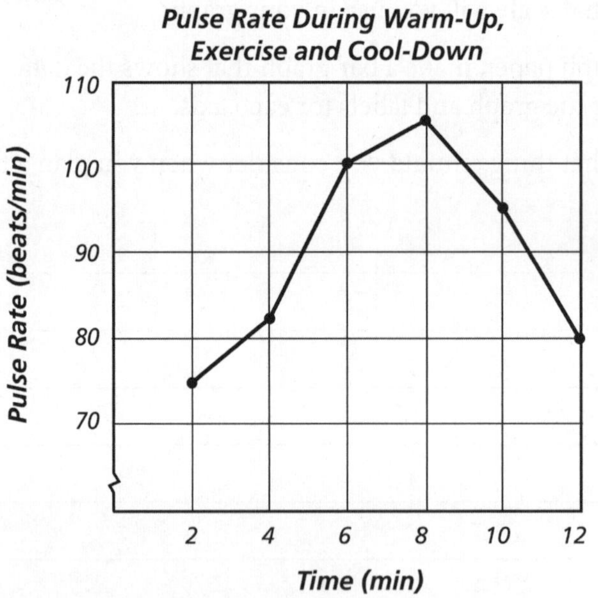

Line graphs show the relationship between two variables. They are useful for showing trends, or patterns, in the data. They can also be used to make inferences and predictions about data you did not directly measure.

Copyright © Pearson Education, Inc., or its affiliates. All Rights Reserved.

Name ______________________ Date ____________ Class ________

Create Line Graphs *(continued)*

What Is a Best Fit Line Graph? On some line graphs, the lines are not drawn from point to point. On the graphs below, the lines are continuous. The lines do not pass through each data point. These graphs are called "best fit graphs." These graphs are used to show trends in data.

The first graph shown below displays a linear relationship. A linear relationship is shown with a straight line. (The word *linear* comes from the word *line.*) This linear graph shows that as the time increases (the manipulated variable), the distance traveled by the car (the responding variable) also increases. You could use this linear graph to make predictions about data that was not actually measured. For example, you might predict that after 120 minutes the car would have traveled 160 km.

The graph in the center shows a curve. You can immediately see that these two variables do not have a linear relationship. This plant grew more quickly in the beginning weeks of the experiment than it did later in the experiment.

The graph on the right shows another type of relationship between variables. Note that the line is a curve that flattens at the top. Populations of living things often show this pattern of growth. The population rises until it reaches a certain size, and then it becomes almost constant.

These line graphs all show the relationship between variables. Using line graphs can help you to show the trend, or pattern, in your experimental data.

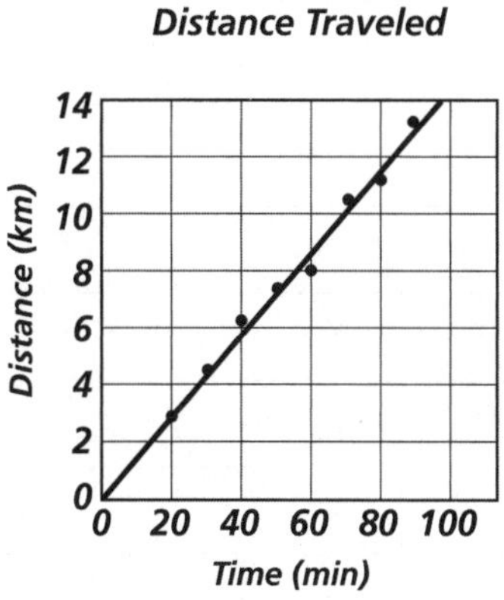

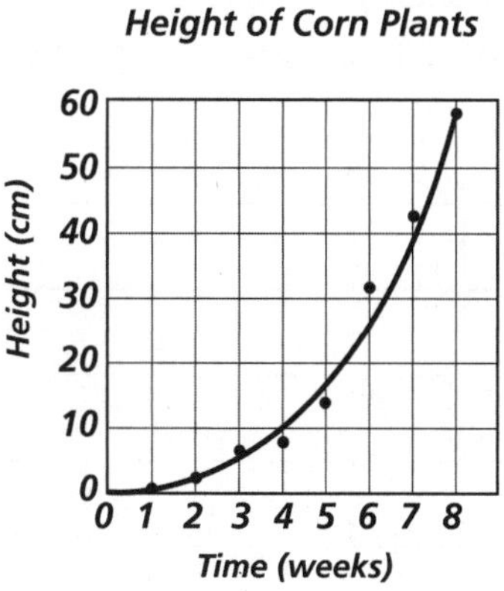

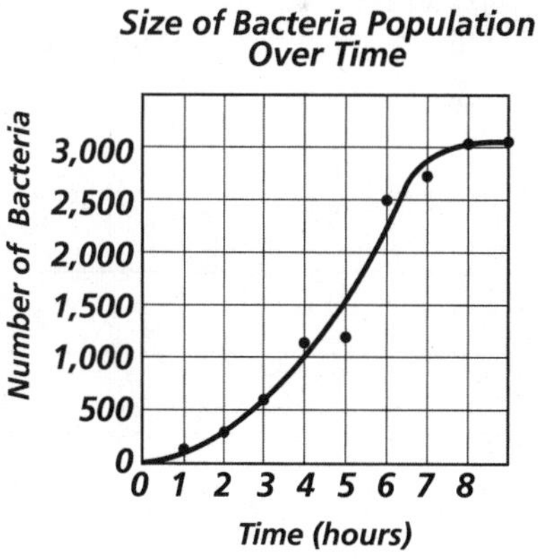

Copyright © Pearson Education, Inc., or its affiliates. All Rights Reserved.

Name ______________________ Date ____________ Class ________

Create Line Graphs *(continued)*

TIPS FOR CREATING LINE GRAPHS

- Draw a horizontal axis and a vertical axis on graph paper. The horizontal axis is called the x-axis, and the vertical axis is called the *y*-axis.
- Label both axes. The horizontal axis should be labeled with the name of the manipulated variable. The vertical axis is labeled with the name of the responding variable. Always include units of measure with your labels.
- Choose a scale for each axis. Determine the least and the greatest value in each set of data to determine the range you will need to represent. Each square of the graph paper should represent the same amount.
- Plot your data points. Follow an imaginary line up from the measurement on the *x*-axis. Then follow a second imaginary line across from the corresponding value on the *y*-axis. The data point is the point at which the two imaginary lines intersect.
- Determine if you should plot from point to point or make a best fit graph. If you are plotting from point to point, use a straightedge to connect adjacent data points. On a best fit graph, the connecting line should be smooth.
- Write a title that clearly describes what is shown in the graph.

Checkpoint Explain how to determine if your experimental data can be shows on a line graph instead of a bar graph. (*Hint*: What characteristics of the manipulated variable lets you know a line graph is appropriate?)

____

____

____

____

____

____

____

____

____

Copyright © Pearson Education, Inc., or its affiliates. All Rights Reserved.

Name ______________________________ Date ______________ Class __________

INQUIRY SKILL FOCUS Practice

Create Line Graphs

Answer the questions that follow on a separate sheet of paper. Then use a sheet of graph paper to make a graph of the data given below.

Earth is surrounded by air. Since air is made of matter, it has mass. Air exerts pressure on Earth's surface. Air pressure is the result of the weight of a column of air pushing down on an area. Air pressure decreases as altitude increases. The air pressure at several different altitudes is shown in the data table.

Air Pressure at Different Altitudes

Altitude (km)	Air Pressure (inches of mercury)
0	30
3	21
6	14
9	9.1
12	5.7
18	2.2

1. The data show the change in air pressure as a result of a change in altitude. What variable should you plot on the horizontal axis?
2. What scale will you use for the vertical axis? How did you choose this scale?
3. On a sheet of graph paper, make a line graph that shows the data. Be sure to include a title for the graph.
4. **Think It Over** Examine the graph to see the trend, or pattern, in this data. Write a sentence that explains the relationship between these two variables.

____

____

____

____

____

____

____

____

Copyright © Pearson Education, Inc., or its affiliates. All Rights Reserved.

Name ______________________ Date ____________ Class ________

INQUIRY SKILL FOCUS Introduction

Create Circle Graphs

Suppose that you and your classmates have chosen topics for your science fair research. The topics are shown in the data table.

Science Fair Research

Topic	Number of Students
Earth Science	3
Biology	1
Chemistry	2
Physics	2

You could show this data in a circle graph. A **circle graph** shows data as parts of a whole. Circle graphs can be used only when you have data for *all* of the categories that make up a given topic. The entire circle represents the total amount. In this case, the circle would represent the eight students doing research for the science fair. The circle is divided into parts, or segments. Each part of the circle graph shows a segment of the whole. In this graph, each segment represents the number of students researching a particular topic.

Science Fair Research

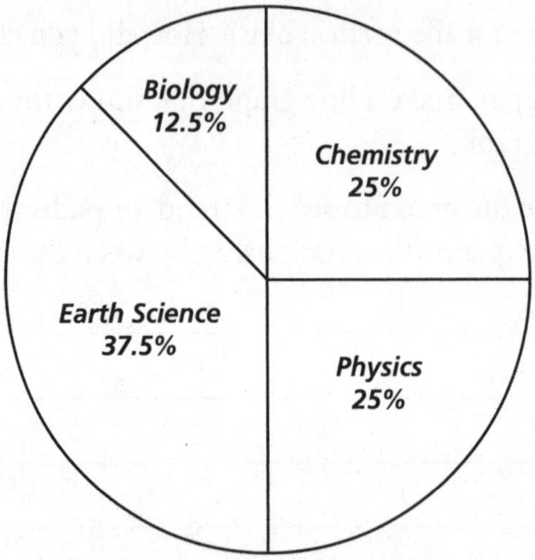

Three of the eight students are doing Earth Science research. The segment representing these students is $\frac{3}{8}$ of the circle. Only one of the eight students is doing Biology research. The segment representing this student is $\frac{1}{8}$ of the circle.

The number of students doing research on a particular topic can be expressed as a percentage. The percentages in a circle graph must add up to 100%.

Copyright © Pearson Education, Inc., or its affiliates. All Rights Reserved.

Name ________________________ Date ______________ Class __________

Create Circle Graphs *(continued)*

TIPS FOR MAKING CIRCLE GRAPHS

Suppose that you have been asked to create a circle graph using information about the grade levels of the students in the science club.

Grade Level of Science Club Members

Grade Level	Number of Club Members
3	12
4	6
5	12
6	18

1. Use your compass to make a circle. Mark the center point and then use a straightedge to draw a line from the center point to the top of the circle.
2. To find the size of the segment that represents science club members who are in third grade, use the following formula. (Remember that a circle contains 360 degrees.)

$$\frac{\text{Number of club members in 3rd grade}}{\text{Total number of club members}} = \frac{x}{\text{Total number of degrees in a circle}}$$

$$\frac{12}{48} = \frac{x}{360°}$$

Cross-multiply, and solve for x. $48x = 12 \times 360°$

$$x = 90°$$

As you can see, the graph representing third graders should contain 90 degrees. The same method is used to find the size of each segment.

3. Use a protractor to measure the angle of the first segment. Use the line you drew to the top of the circle as the 0° line. Draw a line from the center of the circle to the edge to make the first segment. Label the segment.
4. Continue around the circle, measuring and drawing the segments. For each new segment, use the edge of the last segment as the 0° mark. After you have drawn all of the segments, the circle should be full. Make sure you have labeled each segment.

Grade Level	3	4	5	6
Segment Size	90°	45°	90°	135°

5. Find the percentage represented by each segment using the following formula:

$$\frac{\text{Degrees in the segment}}{\text{Degrees in the whole circle}} \times 100\% = \text{Percentage represented by a segment}$$

Checkpoint

Suppose you survey 500 students and find that basketball is the favorite sport of 125 students. What would the angle measurement be in the segment of a circle graph representing 125 students?

____

Copyright © Pearson Education, Inc., or its affiliates. All Rights Reserved.

Name ______________________ Date ______________ Class ________

INQUIRY SKILL FOCUS Practice

Create Circle Graphs

To complete this activity, you will need a compass and a protractor. Use these tools to answer Question 1 on a separate sheet of paper. Answer the remaining questions in the space provided.

Water is used for drinking and washing in homes, for agriculture, in industries, and in the production of power. This data table shows the percent of the total water use used for each of these purposes.

Water Use in the United States

Use	Percentage of Total Water Use
Industries	9
Households and Businesses	10
Agriculture	42
Power Plants	42

1. Make a circle graph to display the data in this table.
2. What two uses combined account for more than 80% of the water use in the United States? Show the individual percentages and the total.
3. **Think It Over** Could you display this data on a bar graph? Why or why not? What does a circle graph show that a bar graph does not?

Copyright © Pearson Education, Inc., or its affiliates. All Rights Reserved.

Name ________________________________ Date ____________ Class __________

INQUIRY SKILL FOCUS Assessment

Skills Test A: Basic Process Skills

Questions 1–4: Read each statement. If the statement is an observation, write "O" on the line at the left. If the statement is an inference, write "I" on the line at the left.

________ **1.** I see snow falling.

________ **2.** Fourteen students are working on the experiment.

________ **3.** The unhealthy plant must not have been given any fertilizer.

________ **4.** The water in the beaker is 22°C.

Questions 5–8: Write the letter of the correct answer on the line at the left.

________ **5.** Which of the following is a qualitative observation that can be made from the illustration at the right?

a. Rain is falling.

b. Trees grow best when it rains regularly.

________ **6.** Which of the following is a quantitative observation that can be made from the illustration?

a. It will probably rain every day next week.

b. Two cm of rain have fallen.

________ **7.** Which of the following is a reasonable inference, based on the illustration?

a. The air temperature is above the freezing temperature.

b. This is an area that gets a large amount of rain each year.

________ **8.** Which of the following statements about observations is true?

a. Observations are information gathered using the senses.

b. Observations are conclusions based on evidence and past experience.

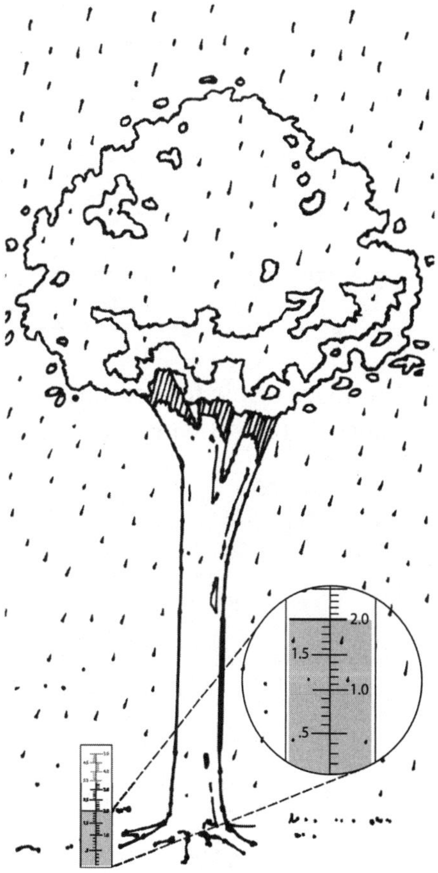

Measurements are in cm.

Copyright © Pearson Education, Inc., or its affiliates. All Rights Reserved.

Name ______________________ Date ____________ Class ________

Skills Test A: Basic Process Skills *(continued)*

Questions 9–10: Write the letter of the correct answer on the line at the left.

_______ **9.** Which of the following is a prediction?

a. The temperature of water can rise or fall depending on the surrounding temperatures.

b. The temperature of the water in the beaker will rise if we heat the beaker.

_______ **10.** Which statement about classification systems is true?

a. There is usually more than one correct way to classify a group of objects.

b. Organizing ideas must be chosen so the groups formed are equal in size.

Questions 11–12: Write your answers in the space provided or on the back of this page.

11. Choose one of the three topics given. Develop an organizing idea that could be used to divide the objects into two groups. Topics: (1) games, (2) clothing, (3) tools used in science class

____

____

____

____

____

____

12. Use the space below to diagram a two-level classification system for one of the topics listed above. If you need more space, use the back of this page.

Copyright © Pearson Education, Inc., or its affiliates. All Rights Reserved.

Name ______________________________ Date ____________ Class ________

Skills Test A: Basic Process Skills *(continued)*

Questions 13–14: Examine the diagram below. Then answer each question by writing the correct answer on the line at the left.

Playground Plan

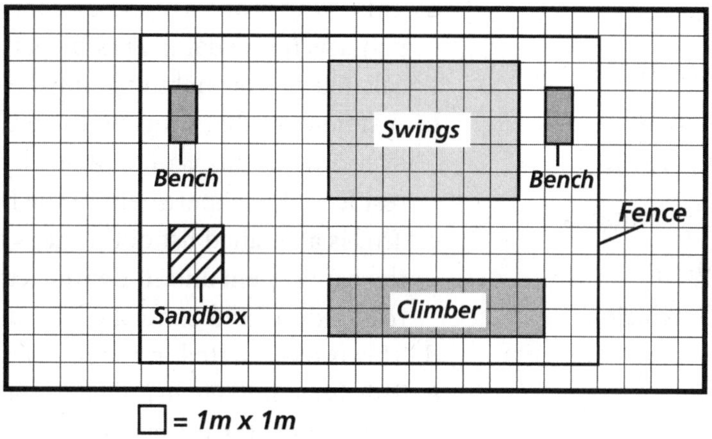

________ **13.** This diagram is an example of a________-dimensional model.

________ **14.** What is the scale of this model?

________ **15.** What would be the actual size of the sandbox shown in this model?

Questions 16–18: Write the letter of the correct answer on the line at the left.

16. Which of the statements about scientific communication is true?

a. Scientific communication should always be free of bias.
b. Scientific communication should not include too many details.

17. When should you include diagrams, charts, and graphs in scientific communications?

a. whenever they would help clarify an idea
b. only when they are easy to understand

18. What is the purpose of recording your procedure in a lab report?

a. to show how you hoped the experiment would work
b. to allow your work to be repeated

Copyright © Pearson Education, Inc., or its affiliates. All Rights Reserved.

Skills Test A: Basic Process Skills *(continued)*

Questions 19–25: Match the correct description from Column 2 with the term in Column 1. Write the letter of the correct answer on the line at the left.

Column 1

________**19.** Question

________**20.** Hypothesis

________**21.** Materials

________**22.** Procedure

________**23.** Observations

________**24.** Analysis

________**25.** Conclusion

Column 2

a. This is the process of looking for trends or patterns in the data, often by performing calculations or making graphs.

b. A statement that sums up the findings of the experiment; this is often made by comparing the results of the experiment to the hypothesis.

c. This is a detailed list of the steps followed during the experiment and it identifies the manipulated, responding, and controlled variables.

d. This is often stated in the *If…then…* form.

e. This is a complete list of all of the items required to complete the experiment.

f. This is the basis of the experiment. It should be specific, about the natural world and free of values or opinions.

g. These are reports of what the senses perceive and measurements made during the experiment. These are often recorded in data tables.

Copyright © Pearson Education, Inc., or its affiliates. All Rights Reserved.

Name ______________________________ Date ______________ Class __________

INQUIRY SKILL FOCUS Assessment

Skills Test B: Measure and Calculate

Questions 1–5: Use the diagram below to answer each question. Write your answer on the line at the left.

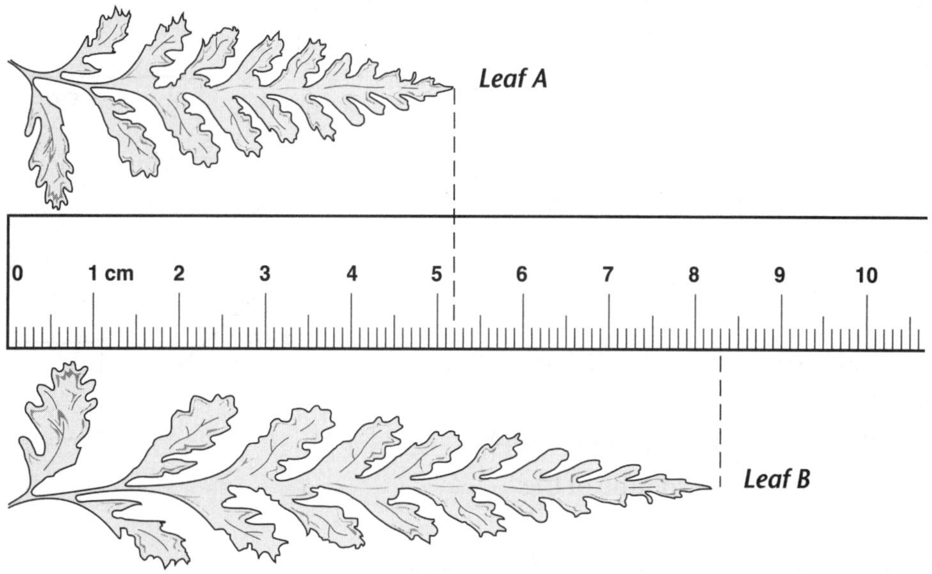

_________ **1.** How many centimeters long is Leaf A?

_________ **2.** How many millimeters long is Leaf A?

_________ **3.** How many centimeters long is Leaf B?

_________ **4.** How many millimeters long is Leaf B?

_________ **5.** How many millimeters longer is Leaf B than Leaf A?

Questions 6–8: Use the diagram below to answer each question. Write your answer on the line at the left.

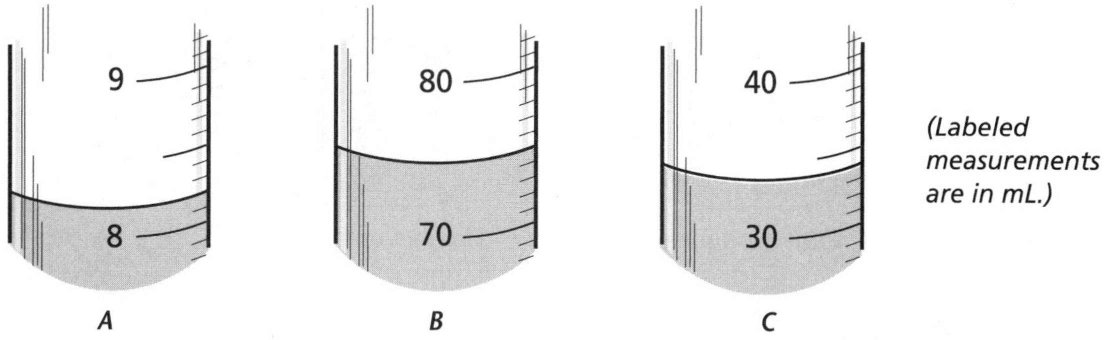

_________ **6.** What is the volume of water in graduated cylinder A?

_________ **7.** What is the volume of water in graduated cylinder B?

_________ **8.** What is the volume of water in graduated cylinder C?

Copyright © Pearson Education, Inc., or its affiliates. All Rights Reserved.

Skills Test B: Measure and Calculate *(continued)*

Questions 9–10: Make the necessary calculations and give your answer in the space below each question.

________ **9.** What is the area of the rectangle on the right?

________ **10.** What is the volume of the figure on the right?

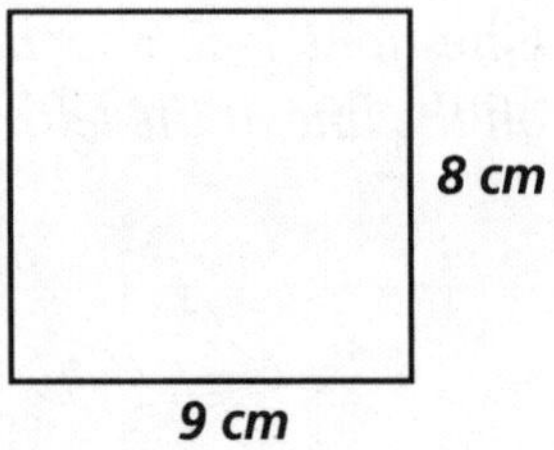

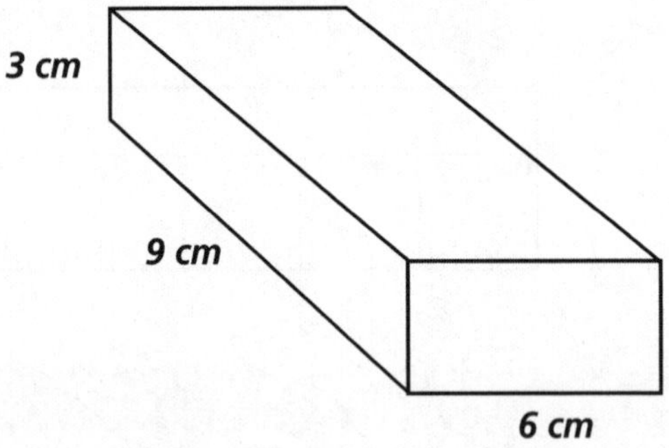

Questions 11–12: Use the information and the diagram provided below to answer the questions. You can show your work below or use the back of this page. Write the answer to each question on the line at the left.

You have decided to measure the volume of a small rock using water and a graduated cylinder. You obtain the results shown below.

________ **11.** What is the volume of the water plus the rock?

________ **12.** What is the volume of the rock alone?

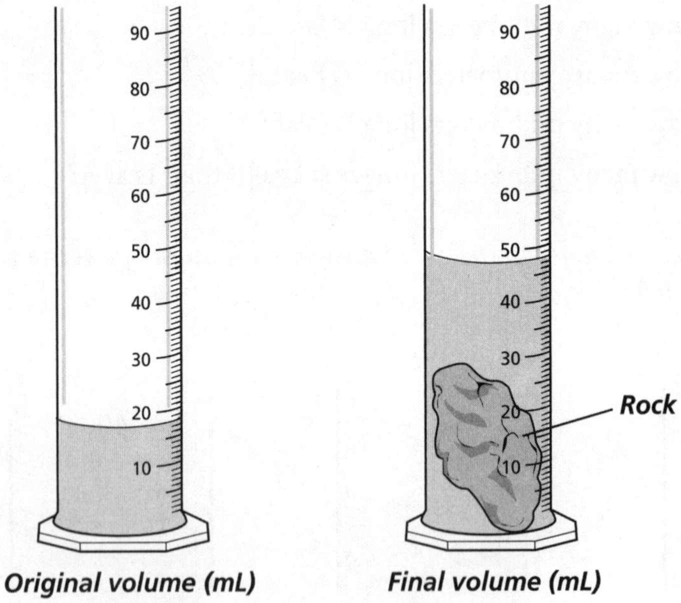

Copyright © Pearson Education, Inc., or its affiliates. All Rights Reserved.

Name ______________________________________ Date______________ Class __________

Skills Test B: Measure and Calculate *(continued)*

Questions 13–14: Use the diagram below to answer each question. Write your answer on the line at the left.

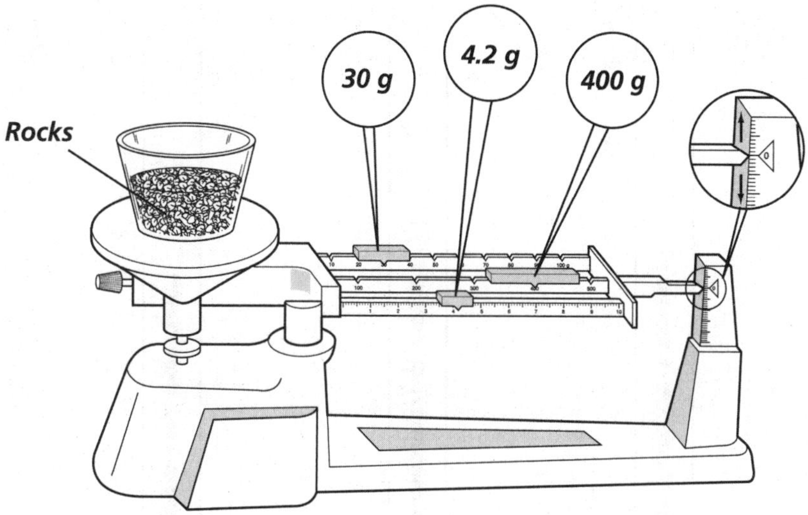

________ **13.** What is the mass of the container and the rocks together?

________ **14.** If the container has a mass of 3.2 grams, what is the mass of the rocks?

Questions 15–19: The diagram below shows five Celsius thermometers. Use these diagrams to answer each question. Write your answer on the line at the left.

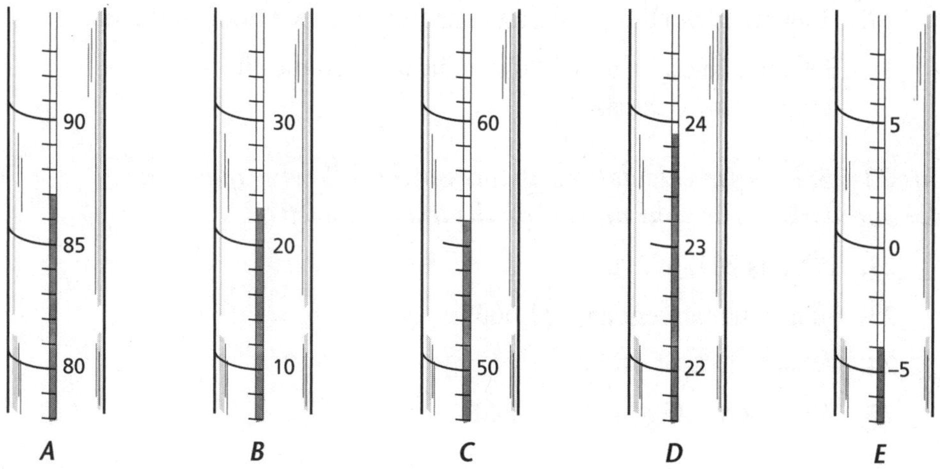

________ **15.** What is the temperature on Thermometer A?

________ **16.** What is the temperature on Thermometer B?

________ **17.** What is the temperature on Thermometer C?

________ **18.** What is the temperature on Thermometer D?

________ **19.** What is the temperature on Thermometer E?

Copyright © Pearson Education, Inc., or its affiliates. All Rights Reserved.

Skills Test B: Measure and Calculate *(continued)*

Questions 20–21: The diagram below shows a Celsius thermometer at different times. Use the diagram to answer each question. Write your answer on the line at the left.

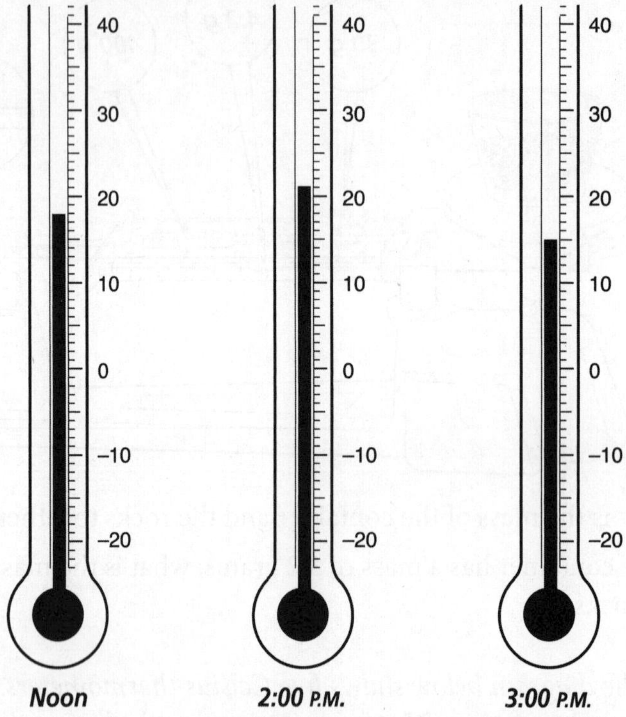

________ **20.** How much did the temperature change between noon and 2 P.M.?

________ **21.** Between 2:00 P.M. and 3:00 P.M., how much did the temperature decrease?

Questions 22–25: Perform calculations to answer the following questions. Show all your work. Write your answer on the line at the left.

________ **22.** What is 20% of 500?

________ **23.** 10 is what percentage of 1,000?

________ **24.** What is 15% of 440?

________ **25.** What percentage of 300 is 240?

Copyright © Pearson Education, Inc., or its affiliates. All Rights Reserved.

Name ______________________________ Date ____________ Class ________

INQUIRY SKILL FOCUS Assessment

Skills Test C: Design Experiments

Questions 1–5: Write the letter of the correct answer on the line at the left.

________ **1.** *If water contains sugar, then it will freeze at a lower temperature than fresh water does.* This is an example of a(n)
a. scientific question.
b. hypothesis.
c. operational definition.
d. procedure.

________ **2.** Measurements gathered during an experiment are an example of
a. operational definitions.
b. inferences.
c. variables.
d. observations.

________ **3.** What is the manipulated variable in an experiment to determine whether the distance a skateboard rolls is affected by the size of the skateboard's wheels?
a. the surface on which the skateboard rolls
b. the force with which the skateboard is pushed
c. the size of the skateboard's wheels
d. the distance the skateboard rolls

________ **4.** Which of the following is a detailed list of the steps in an experiment?
a. conclusion
b. operational definition
c. procedure
d. observations

________ **5.** During an experiment, what do you record in a data table?
a. the observations
b. the procedure
c. the conclusion
d. the hypothesis

Copyright © Pearson Education, Inc., or its affiliates. All Rights Reserved.

Name ______________________________ Date ______________ Class __________

Skills Test C: Design Experiments *(continued)*

Questions 6–12: Write the letter of the correct answer on the line at the left.

________ **6.** Which statement about scientific questions is true?
a. Scientific questions should avoid bias.
b. Scientific questions should be as general as possible.

________ **7.** Which statement about the variables in an experiment is true?
a. Controlled variables can change during the experiment.
b. The manipulated variable is changed intentionally during the experiment.

________ **8.** Which of the following is a scientific question?
a. Which kinds of metal are attracted by a magnet?
b. Which kinds of metal make the most attractive jewelry?

________ **9.** Which of these terms, when used in an experiment, would require an operational definition?
a. a steep ramp
b. a ramp at a 45° angle

________ **10.** Which of these would be most useful for the process of interpreting data?
a. a microscope
b. a graph

________ **11.** What is the manipulated variable in an experiment to determine whether the use of antibacterial soap changes the average number of student absences from school?
a. use of regular or antibacterial soap
b. number of student absences from school

________ **12.** What is the purpose of a control group in an experiment?
a. to serve as a standard of comparison
b. to be the group exposed to changes in the manipulated variable

Copyright © Pearson Education, Inc., or its affiliates. All Rights Reserved.

Name ______________________________ Date ____________ Class __________

Skills Test C: Design Experiments *(continued)*

Questions 13–20: Column 1 lists the key components of an experiment. Column 2 presents examples of these components.Match each term in Column 1 with the correct statement from Column 2.Write the letter of the correct answer on the line at the left.

Column 1

________ **13.** controlled variables

________ **14.** hypothesis

________ **15.** scientific question

________ **16.** operational definition

________ **17.** conclusion

________ **18.** data

________ **19.** manipulated variable

________ **20.** responding variable

Column 2

a. If plants receive fertilizer, they will make more beautiful bouquets.

b. If a box is pushed with a greater force, then it will move a greater distance.

c. Does hot water freeze more quickly than cold water?

d. The pulse rates measured were 72, 68, and 83 beats per minute.

e. Water was considered cold when it was 10°C or less.

f. Plants received 2 g, 4 g, 6 g, or 8 g of fertilizer weekly.

g. All of the toy cars were rolled over the same surface and were pushed with the same amount of force.

h. The rate of plant growth in an experiment to measure the affect of watering on rate of growth.

i. Plants that received Fertilizer A grew 2 cm per week more than plants with no fertilizer. Fertilizer A makes plants grow faster.

Copyright © Pearson Education, Inc., or its affiliates. All Rights Reserved.

Name ______________________________ Date ______________ Class __________

Skills Test C: Design Experiments *(continued)*

Questions 21–25. Read the description of the experiment. Then write the letter of the terms from the list under the correct description.

We did an experiment to see if the distance a toy car rolled was affected by the size of the car's wheels. First, we set up a ramp at 45°. Each of the toy cars was rolled down the same ramp. At the bottom of the ramp, the cars rolled onto the classroom floor. We used a metric ruler to measure the distance each car rolled. The cars we used all had the same mass, but the wheels on each car were a different diameter. Our hypothesis was: If a toy car has larger wheels, then it will roll a greater distance.

a. masses of the toy cars

b. distance traveled by the toy cars

c. height of the ramp

d. surface at the bottom of the ramp

e. size of the toy car's wheels

Controlled Variables	Manipulated Variable	Responding Variable
21. ____________	**24.** ____________	**25.** ____________
22. ____________		
23. ____________		
____________		

Copyright © Pearson Education, Inc., or its affiliates. All Rights Reserved.

Name ______________________ Date ____________ Class __________

INQUIRY SKILL FOCUS Assessment

Skills Test D: Data Tables and Graphs

Questions 1–6: Write the letter of the correct answer on the line at the left.

_______ **1.** When should a data table for an experiment be made?
- **a.** while the experiment is being performed
- **b.** after all of the data have been collected
- **c.** when you are ready to make conclusions
- **d.** while you are planning the experiment

_______ **2.** In an experiment to show the change in a variable over time, what type of graph are you most likely to use to show trends in the data?
- **a.** bar graph
- **b.** line graph
- **c.** double bar graph
- **d.** circle graph

_______ **3.** What does each segment in a circle graph represent?
- **a.** a responding variable
- **b.** a portion of a whole
- **c.** a controlled variable
- **d.** a rate of change

_______ **4.** What is usually represented on the vertical axis of a line graph?
- **a.** the controlled variables
- **b.** the responding variable
- **c.** the size of the control group
- **d.** the manipulated variable

_______ **5.** What do the bars in a bar graph represent?
- **a.** separate, but related categories
- **b.** the change in a variable over time
- **c.** the continuous variable
- **d.** unrelated categories

_______ **6.** What should a data table include?
- **a.** spaces for the best data from the experiment
- **b.** spaces for measurements from the first trial of an experiment
- **c.** spaces for all of the data that are collected during the experiment
- **d.** spaces for the data that best supports the hypothesis

Copyright © Pearson Education, Inc., or its affiliates. All Rights Reserved.

Name ______________________ Date ______________ Class __________

Skills Test D: Data Tables and Graphs *(continued)*

Question 7: Use a separate sheet of paper to make a data table that the students could use during their experiment.

7. Marcus and his lab group were studying the rate at which different liquids flowed down a ramp. They followed the procedure described below. Create a data table the lab group could use to record their data. Include a title and labels for the columns and rows.

> In this experiment we will measure the rate at which several different kinds of liquids flow down a ramp. We will set up a ramp to use for each trial. For each type of liquid we will run two trials and average the results of the two trials. We will test the following liquids: water, dish soap, and honey. We will measure the time in seconds that it takes the liquid to travel from the top of the ramp to reach the bottom of the ramp.

Fruit and Vegetable Consumption Survey Results

Average Number of Servings Per Day	Number of Students
4	40
5	100
6	60
7	100
8	100

Question 8: Use the circle below to make your circle graph.

8. While studying health and nutrition in science class, Julia's lab group decided to survey students in their school to determine how many servings of fruits and vegetables the students ate daily. They listed their results in the data table above. Use the data to make a circle graph. Label each segment and add a title.

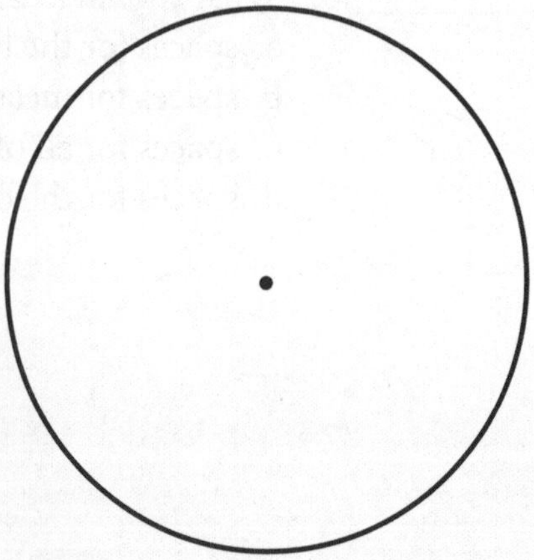

Copyright © Pearson Education, Inc., or its affiliates. All Rights Reserved.

Skills Test D: Data Tables and Graphs *(continued)*

Question 9: Use the graph paper below to make your graph. Be sure to label the axes and include a title.

9. When Keiko's science class studied geology, she and several classmates ordered rock collections from different suppliers. The class then classified the rocks as igneous, sedimentary, or metamorphic, and made a data table of the results. Use the information from the table to create a bar graph.

Origin of Collected Rocks

Type of Rock	Number of Rocks
Igneous	38
Sedimentary	52
Metamorphic	47

Copyright © Pearson Education, Inc., or its affiliates. All Rights Reserved.

Skills Test D: Data Tables and Graphs *(continued)*

Question 10: Use the graph paper below to make your graph. Be sure to label the axes and include a title.

10. One day while their class was studying weather, Marco and Emily recorded the outside temperature every 20 minutes for several hours. Their results appear in the data table. Use that information to create a line graph.

Outside Temperature

Time (min)	Temperature (°C)
0	12
20	13
40	13
60	15
80	16
100	16
120	18
140	18
160	17
180	18

Copyright © Pearson Education, Inc., or its affiliates. All Rights Reserved.

Observe

Introduction *(pp. 1–2)*

Checkpoint: Answers will vary. Sample: The desk is brown. (qualitative) The desk has 4 legs. (quantitative)

Practice *(p. 3)*

1. Sample: There is liquid in the graduated cylinders. There is liquid in the beaker. There are bubbles in the liquid in the beaker. The liquids are different colors.
2. Sample: There are about 400 mL of liquid in the beaker. There are 2 graduated cylinders. There is 350 mL of liquid in cylinder A.
3. Inference. Explanations will vary. Sample: This is an inference, because I can't be sure that the beaker contains a mixture of liquids from the graduated cylinders.
4. Observation. Explanations will vary. Sample: I can use my sense of sight to determine the colors of the liquids, so this is an observation.
5. Inference. Explanations will vary. Sample: This is an inference, because I did not observe the liquids being mixed.
6. Sample: An observation must be based directly on the senses such as sight or hearing. Example: A cat walked into the room. A statement explaining an observation is an inference. Example: The cat is looking for food.

Infer

Introduction *(pp. 4–5)*

Checkpoint: Answers will vary. Sample: I would infer that this car had a flat tire. I can't be sure this inference is correct, because the tire has been removed, so I can't observe it directly.

Practice *(pp. 6–7)*

1. bird 1
2. Sample: A beak that is adapted to get nectar from flowers would be long and slender.
3. bird 4
4. bird 2
5. bird 1
6. bird 5
7. bird 3
8. Sample: I observed that bird 3 has a webbed foot that would aid in swimming.
9. Information about the bird's diet and feet would allow me to confirm my inferences. I could get this information by observing birds in their natural habitat or by using reference materials.

Predict

Introduction *(pp. 8–9)*

Checkpoint: Answers will vary. Sample: I predict that fertilizer will make a plant grow more quickly than a plant that does not receive fertilizer. I would test this by comparing two like plants in identical conditions, except that one plant gets fertilizer and the other does not.

Practice *(pp. 10–11)*

1. Sample: the solubility of all three substances increases with temperature; the solubility of Substance C is increasing more quickly than the solubility of Substance A.
2. Substance B
3. I can predict unmeasured data by extending the lines in the graph. The solubility of Substance A at 35°C would be about 23 g. I predict that about 34 grams of B will dissolve at 35°C.

Copyright © Pearson Education, Inc., or its affiliates. All Rights Reserved.

4. These substances will not have the same solubilities at any temperature below 25°.
5. Although the graph lines for Substances A and C are both increasing, the solubility of Substance C is increasing at a greater rate then that of Substance A, so the solubilities will not be the same over 20°.
6. No, I can't be sure these predictions are correct. I could check the solubility of these substances at 35°C to see if the predictions are correct.

Classify

Introduction *(pp.12–13)*

Checkpoint: Answers will vary. Sample: Animals can be classified by whether or not they have a backbone, by where they live, or by the type of foods they eat.

Practice *(pp. 14–15)*

1. Sample: widely used/scarcely used; limited supply/unlimited supply; from a living source/from a nonliving source
2. Sample: renewable/nonrenewable/inexhaustible
3. Answers will vary depending upon the organizing ideas used.
4. Sample: I could use the following organizing ideas: resources that create lots of pollution when used/"clean" energy sources; resources that harm the environment when they are obtained or mined/resources that can be obtained without damaging Earth.

Make Models

Introduction *(p. 16)*

Checkpoint: Answers will vary. Sample: Maps and globes are similar in that they both are scale models. They are different because globes are three-dimensional, and maps are two-dimensional.

Practice *(p. 17)*

1. Answers will vary. Sample: I will use the scale 1 cm = 25 m, so the layers of the atmosphere will fit on the provided page.
2. Sample: The model will be 22 cm from bottom to top.
3. Sample: I had trouble showing the troposhere because it's only 12 kg deep. The exosphere has no definite outer limit, it extends forever into space. This would be impossible to show on a scale model.

   Students' models should accurately reflect the information on page 17. Each model should include a scale.

Communicate

Introduction *(pp. 19–20)*

Checkpoint: Answers will vary. Sample: The purpose of both forms of communication is to share ideas with others. Both written and oral communication can use graphs, tables, charts, and diagrams. Written communications can be examined at any time. Oral communications allow researchers to interact immediately.

Practice *(pp. 21–22)*

1. Sample: Does exercise change a person's pulse rate?
2. Sample: If I exercise, then my pulse rate will increase.
3. Sample: watch with a second hand, an area with stairs, a large space in which to do jumping jacks
4. Sample: Step 1: measure pulse in a 15-second period while resting; Step 2: measure pulse in a 15-second period after one minute of jumping jacks; Step 3: measure pulse in a 15-second period after one minute of stair climbing; Step 4: measure pulse in a 15-second period five minutes after exercise

Copyright © Pearson Education, Inc., or its affiliates. All Rights Reserved.

5. Data tables should use the provided data and format.
6. Sample: Calculations should show the 15-second pulse rate numbers multiplied by 4 to find pulse rate per minute.
7. Sample: Pulse rate increases during exercise, and returns almost to normal within 5 minutes.
8. Sample: Does pulse rate continue to increase for the entire length of time exercise is performed? This could be investigated by measuring pulse rate at one-minute intervals during an extended period of exercise.

Measure

Introduction *(pp. 23–24)*

Checkpoint: 420,000 m; 420 km × 1,000 meters per kilometer = 420,000 m

Practice *(pp. 25–26)*

1. 31 mm
2. 3.1 cm
3. 82 mm
4. 8.2 cm
5. 59 mm
6. 5.9 cm
7. 5.1 cm
8. length 5 cm, width 3 cm
9. 15 cm^2
10. length 9 mm, width 4 mm, height 4 mm
11. 144 mm^3
12. Sample: First I would convert centimeters to millimeters so all of the measurements have the same units. 5 cm = 50 mm. Then I would multiply to find the area. 50 mm × 32 mm = 1,600 mm^2. (16 cm^2 if student works with centimeters)

Practice *(p. 27)*

1. 15 mL
2. 48 mL
3. 23 mL
4. 18 mL
5. 42 mL
6. 24 mL
7. Sample: First I would note the numbered markings and determine the gap between those numbers. Then I would see how many spaces appear in this gap, and then divide to find what each marking stands for. For example, 10 mL divided by 5 spaces means each line indicates 2 mL.

Practice *(p. 28)*

1. 143.5 g
2. 115 g
3. 52,000 milligrams.

Practice *(p. 29)*

1. 14°C
2. –5°C
3. 32°C
4. 31°C
5. 27.5°C
6. The change was 8°C. From 7°C to zero, the temperature change was 7 degrees, then it dropped on more degree from 0°C to –1°C, for a total drop of 8 degrees.

Calculate

Introduction *(pp. 30–31)*

Checkpoint: 250 heads, 250 tails

Practice *(p. 32)*

1. 20%
2. 75%

Copyright © Pearson Education, Inc., or its affiliates. All Rights Reserved.

3. 110 calories
4. 187.5 joules
5. Sample: I can calculate 50% by simply taking half of a number.

Design an Experiment

Introduction *(pp. 33–34)*

Checkpoint: No, an experiment cannot have two manipulated variables. If there is more than one manipulated variable, it will not be possible to determine what has caused any changes in the responding variable.

Pose Questions

Introduction *(p. 35)*

Checkpoint: Sample: Does the amount of light a plant receives affect its growth rate? I would investigate this by measuring the growth rate of several plants that receive different amounts of light, but are treated identically in every other respect.

Practice *(p. 36)*

1. Yes; does the rater of water flow affect the rate of erosion?
2. no
3. no
4. Yes; does the composition of soil affect the growth rate of plants?
5. Yes; at what rate do different rock types break down to form soil?
6. Sample: What insects do the birds in this area eat? What animals in this area are most important in dispersing the seeds of the blackberry plants?
7. Sample: How does the root structure of desert plants differ from the root structure of aquatic plants? How does the leaf shape of maple trees and oak trees differ?
8. Opinion and personal values cannot be tested by observing or measuring the natural world, so they can't be answered scientifically.

Develop a Hypothesis

Introduction *(p. 37)*

Checkpoint: Sample: Scientific question: Does the temperature affect the length of time it takes for a seed to germinate? Hypothesis: If the temperature is 25°C, then germination will occur more quickly than at other temperatures.

Practice *(pp. 38–39)*

1. Sample: Does the size of a toy car's wheels affect how far it rolls?
2. Sample: If a toy car has larger wheels, then it will roll farther.
3. Sample: I would use a ramp setup like the one shown on the opposite page to measure how far each car rolls. I would use cars with several different sizes of wheels, and I would control all of the other variables.
4. Sample: One car may have a greater mass than the other, and that might affect how far it rolls. One car might have less internal friction than the other, and that might affect how far it rolls.
5. Sample: If a car has a greater mass, then it will roll a greater distance. If a car has less friction, then it will roll a greater distance.
6. No, a hypothesis is a possible explanation for a set of observations. It is not necessarily correct. An investigation can be used to show whether or not a hypothesis is correct.

Control Variables

Introduction *(pp. 40–41)*

Checkpoint: Sample: amount of light shining per unit time, the length of the

Copyright © Pearson Education, Inc., or its affiliates. All Rights Reserved.

light period, amount of water the plants receive, the type of soil the plants are in, the temperature of the room

Practice (p. 42)

1. manipulated: temperature; responding: rate of ripening
2. manipulated : mass of car; responding: rate of rolling
3. Sample: the surface over which the cars are traveling, the force with which the cars are pushed, the size of the cars' wheels
4. Sample: All participants would have to be in good health when the experiment started. I would have a control group that washed with regular soap report how many times they were sick in one month, and an experimental group that used antibacterial soap report how many times they were sick in one month.
5. The number of times people get sick depends on many factors. By using a large group, you can see the effect of the soaps, rather than seeing possible effects of individual health differences.
6. No, if an experiment has two manipulated variables, there is no way to know what caused any changes in the responding variable. This student could not draw valid conclusions from her work.

Form Operational Definitions

Introduction (pp. 43–44)

Checkpoint: This is incorrect because anyone trying to answer a scientific question may need to repeat his or her work or the work of others. Also, within a single experiment, all observations should be made in exactly the same way.

Practice (p. 45)

1. Sample: at a rate of 2 or more cm per week
2. Sample: able to hold the greatest mass without ripping
3. Sample: when the wind speed is 35 km per hour or more
4. Sample: use a Celsius thermometer to measure the temperature in degrees Celsius, and record the temperature every minute
5. Sample: 0.5 meter or longer
6. I would need to make a new operational definition for the term long legs, because in general, horses have longer legs than even large dogs. Certain terms can have different operational definitions in different experiments because the purposes of similar experiments can be very different; experimental subjects, materials, and tools can also be different.

Interpret Data

Introduction (pp. 46–47)

Checkpoint: Sample: Your knowledge about the topic may be incomplete or incorrect. There may have been a mistake in the experiment, causing the data to be incorrect. There may have been a mistake in the interpretation of the data, which would cause your inferences not to match your pervious knowledge of a topic.

Practice (p. 48)

1. Sample: This graph shows the rate of cooling of water in an insulated and a noninsulated can over the course of one hour.
2. 0 min, 55 min, 60 min
3. Any time between 10 min and 20 min
4. The water in the noninsulated container cooled more quickly than the water in the insulated container.

Copyright © Pearson Education, Inc., or its affiliates. All Rights Reserved.

5. 20°C; the water has probably reched the temperature of the room
6. Yes, since insulation slows temperature changes, it would make sense that water in an insulated container would cool more slowly than water in a noninsulated container.

Draw Conclusions

Introduction *(pp. 49–50)*

Checkpoint: No, you should never change your data so that your conclusion shows your hypothesis is correct. Your conclusion should always accurately reflect the experimental data.

Practice *(p. 51)*

1. Sample: Elodea grown in chlorinated water grows more slowly than elodea in water with no chlorine.
2. Correct
3. Elodea grown in chlorinated water grows more slowly than elodea grown in water with no chlorine.
4. This conclusion can't be logically drawn because the experiment tested only elodea, not all aquarium plants.
5. Sample: Yes, even if your hypothesis is shown to be incorrect, you would have still learned that elodea grow better without chlorine in the water.

Design an Experiment

Practice *(p. 52)*

You may wish to arrange students in small groups to help each other improve on their experimental designs, and then discuss selected designs as a whole class.

1–10. Students should follow all the relevant guidelines in their experimental designs.

11. Operational definitions are an important part of an experimental design because they allow the work to be repeated by myself or by others.

Create Data Tables

Introduction *(pp. 53–54)*

Checkpoint: The manipulated and responding variables need to be observed and measured during the experiment, and these measurements and observations must be recorded in an organized way.

Practice *(pp. 55–56)*

1. Manipulated: moisture; Responding: area of mold
2. Once per day for 5 days
3. Area: cm^2
4. The table should include a title, a column for Time (days), one column each for the dry bread and wet bread, and rows for measurements from all 5 days. The label cm^2 should appear in an appropriate locations.
5. 3 trials
6. Averaging the results of the three trials
7. Once for each trial
8. Data tables should include a title, on column each for rock salt and table salt, and rows for all 3 trials and the averages.
9. Numbers without unit labels have no meaning. Units allow others to interpret your data and can be used when comparing your work to the work of others.

Create Bar Graphs

Note: Bar graphs can be created with the bars running horizontally or vertically. However, many students have difficulty deciding which axis to use for each

Copyright © Pearson Education, Inc., or its affiliates. All Rights Reserved.

variable, so this book introduces only one approach. Once your students are adept at graphing, you many want to allow either type of bar graph.

Introduction *(pp. 57–58)*

Checkpoint: The scale needs to be chosen so both the largest and the smallest measurements can be included on the graph.

Practice *(p. 59)*

1. Speed of Sound (m/s)
2. Sample: one block = 500 m/s
3. The bars for speed will get taller from left to right.
4. You need to consider what title will briefly describe the contents of the graph, so others know what data the graph contains.

Create Line Graphs

Introduction *(pp. 60–62)*

Checkpoint: Experimental data can be shown on a line graph if the manipulated variable is continuous, that is, it can have measurements in between those already taken.

Practice *(p. 63)*

1. change in altitude
2. Sample: One square = 2 inches of mercury, because the smallest measurement is just slightly more than this scale, and the largest measurement will fit on the graph.
3. The graph should show all data accurately, be completely labelled, and have a title.
4. Air pressure decreases as altitude increases.

Create Circle Graphs

Introduction *(pp. 64–65)*

Checkpoint: 90°

Practice *(p. 66)*

1. The segments of the circle graph should have the following angle measurements:
   Industries and Mining: 32.4°
   Households and Business: 36°
   Agriculture: 151.2°
   Power Plants: 140.4°
2. Agriculture (42%) and power plants (39%), account for a total of 81% of the water use.
3. Yes, these are separate, but related, categories, and they could be shown on a bar graph. A circle graph shows the percentage of the total amount used by each category. This is not shown on a bar graph.

Test A: Basic Process Skills

(pp. 67–70)

1. O
2. O
3. I
4. O
5. a
6. b
7. a
8. a
9. b
10. a
11. Answers will vary, but the idea must be chosen so two groups are formed.
12. Answers will vary, but they must include two correct levels of classification.
13. two
14. one square = 1 m by 1 m
15. 2 meters × 2 meters
16. a
17. a
18. b

Copyright © Pearson Education, Inc., or its affiliates. All Rights Reserved.

19. f
20. d
21. e
22. c
23. g
24. a
25. b

Test B: Measure and Calculate *(pp. 71–74)*

1. 5.2 cm
2. 52 mm
3. 8.3 cm
4. 83 mm
5. 31 mm
6. 8.2 mL
7. 75 mL
8. 34 mL
9. 72 cm2
10. 162 cm3
11. 48 mL
12. 30 mL
13. 534.2 g
14. 431 g
15. 87°C
16. 23°C
17. 56°C
18. 23.9°C
19. −4°C
20. 3°C
21. 6°C
22. 100
23. 1%
24. 66
25. 80%

Test C: Design Experiments *(pp. 75–78)*

1. b
2. d
3. c
4. c
5. a
6. a
7. b
8. a
9. a
10. b
11. a
12. a
13. g
14. b
15. c
16. e
17. i
18. d
19. f
20. h

21–23. a, c, d (in any order)

24. e
25. b

Test D: Data Tables and Graphs *(pp. 79–82)*

1. d
2. b
3. b
4. b
5. a
6. c

Copyright © Pearson Education, Inc., or its affiliates. All Rights Reserved.

7. Data tables should include space for the results from two trials on three liquids. There should also be a space for the average results.

Partial credit could be given as follows:

(6 points) Using the correct variables for column an/or row heads.

(2 points) Using appropriate units in heads.

(2 points) Giving the table an appropriate title.

8. Segment measurements for the circle graph should be

   5 = 90°

   7 = 90°

   8 = 90°

   4 = 30°

   6 = 60°

Partial credit could be given as follows:

(6 points) Calculating the correct number of degrees in each segment.

(6 points) Estimating and drawing the segments on the circle template.

(2 points) Labeling each segment appropriately.

(1 point) Using an appropriate title.

9. The completed bar graph should resemble the one below.

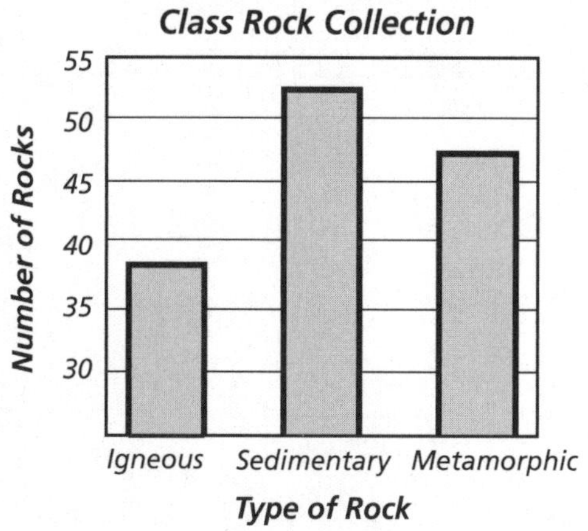

Partial credit could be given as follows:

(4 points) Drawing horizontal and vertical axes.

(5 points) Choosing and labeling categories for each axis.

(5 points) Determining an appropriate, uniform scale.

(5 points) Labeling the scale at regular intervals.

(1 point) Writing a title for the graph.

Copyright © Pearson Education, Inc., or its affiliates. All Rights Reserved.

10. The completed line graph should resemble the one at the bottom of the page.

Partial credit could be given as follows:

(3 points) Drawing horizontal and vertical axes.

(3 points) Labeling the axes correctly.

(3 points) Creating appropriate scales for the axes.

(2 points) Marking off equally spaced intervals along each axis.

(3 points) Correctly plotting and connecting the points.

(1 point) Giving the graph an appropriate title.

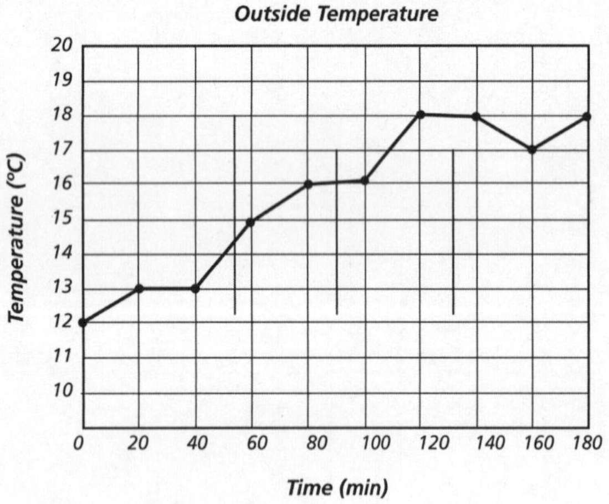

Copyright © Pearson Education, Inc., or its affiliates. All Rights Reserved.